11th International Conference on 3D Radiation Dosimetry (IC3DDose)

Journal of Physics: Conference Series Volume 2167

Quebec City, Canada
15 June 2021

ISBN: 978-1-7138-5138-7
ISSN: 1742-6588

Printed from e-media with permission by:

Curran Associates, Inc.
57 Morehouse Lane
Red Hook, NY 12571

Some format issues inherent in the e-media version may also appear in this print version.

Institute of Physics
Dirac House, Temple Back
Bristol BS1 6BE UK

Phone: 44 1 17 929 7481
Fax: 44 1 17 920 0979

techtracking@iop.org

Additional copies of this publication are available from:

Curran Associates, Inc.
57 Morehouse Lane
Red Hook, NY 12571 USA
Phone: 845-758-0400
Fax: 845-758-2633
Email: curran@proceedings.com
Web: www.proceedings.com

TABLE OF CONTENTS

Journal of Physics: Conference Series **2167** (2022) 011001 doi:10.1088/1742-6596/2167/1/011001

IC3Ddose 2021 - 11th International Conference on 3D and Advanced Dosimetry

Preface

The 11th edition of the International Conference on Three Dimensional (3D) and Advanced Dosimetry, IC3Ddose, was scheduled to meet in Quebec City, Quebec, Canada in 2020. The conference, initiated originally in Kentucky in 1999 and then continued roughly every two years thereafter in venues across the globe, was set for a Canadian venue.

By late February 2020 the organizers had received over 60 submissions for proffered presentations to the conference and had recruited a strong cohort of 16 invited speakers. The submissions covered topics from film dosimetry, 3D dosimetry with chemical systems, EPID dosimetry, scintillation and Cherenkov based techniques, challenges of clinical dose delivery validation with dynamic techniques and small radiation fields, end-to-end QA, and more. All was ready for a very successful meeting in June 2020, and then Covid-19 stopped all life as usual in March of that year.

In late 2020 it became clear that it would not be possible to just postpone the 2020 meeting to 2021, so it was decided to turn IC3Ddose021 into a virtual meeting. This enabled those who had submitted work for the conference to still present their research albeit remotely in a meeting from May 10th to 13th, 2021. Some of the original submissions were withdrawn because they had been already published in the interim or because authors could not attend the meeting. Thirty-eight papers were presented in 21 oral presentations and 17 posters over the four days of IC3Ddose21. The sessions were scheduled to accommodate 93 participants from 13 countries across multiple time-zones. The oral sessions over the meetings first three days were given over the *Zoom* remote platform that everyone had become proficient with in the last year. To replicate the relaxed personal interactions long fostered in IC3Ddose, interactive sessions after each day's oral presentation with poster viewing, vendor exhibits and informal gathering rooms for open discussion were moved to an interactive virtual platform called *Spatial.Chat*.

Five commercial vendors also participated in the meeting, each had two rooms in the virtual conference space, one for a booth and one to show videos highlighting their products and services. We thank these partners for their support to make this virtual meeting a reality: Elekta, GeVero Co, MedScint, Modus QA, and RTsafe.

As the local organizing committee, we would like to also thank all the participants for the excellent science presented during the formal sessions and for the very interesting discussions in the networking sessions afterwards. It was encouraging in this virtual meeting to see many attendees in the discussion spaces long after the presentations had finished. We are looking forward for the June 2022 meeting in full 3D!

L. John Schreiner, Louis Archambault and Luc Beaulieu

International Scientific Committee

Louis Archambault, Canada; Sven Bäck, Sweden; Clive Baldock, Australia; Luc Beaulieu, Canada (chair); Sam Beddar, USA; Yves de Deene, Australia; Simon Doran, UK; Geoffrey Ibbott, USA; Andrew Jirasek, Canada; Kevin Jordan, Canada; Martin Lepage, Canada; Boyd McCurdy, Canada; Mark Oldham, USA; Brian W Pogue, USA; L. John Schreiner, Canada; Hiroshi Yasuda, Japan

Local Organizing Committee

Louis Archambault, Canada; Luc Beaulieu, Canada; L. John Schreiner, Canada;

3D DOSE
Québec - 2020

Preliminary Program of the 2021 virtual **INTERNATIONAL CONFERENCE ON 3D and ADVANCED DOSIMETRY**, May 10-13 2021.

At a glance

Time	Monday May 10th	Tuesday May 11th	Wednesday May 12th	Thursday May 13th
08h-09h30 EDT 14h-15h30 CE 21h-22h30 JPN 22h-23h30 AUS	Rising Star Competition			Poster session + networking
09h30-10h30 EDT 15h30-16h30 CET 22h30-23h30 JPN 23h30-00h30 AUS	Networking session			
10h00 - 11h30 EDT 16h00 - 17h30 CET 23h00-00h30 JPN 00h00-1h30 AUS			Contributed oral presentations Session 2	
11h30-12h30 EDT 17h30-18h30 CET 00h30-01h30 JPN 01h30-02h30 AUS			Networking session	
18h00-19h30 EDT 00h00-01h30 CET (+1d) 07h00-08h30 JPN (+1d) 08h00-09h30 AUS		Contributed oral presentations Session 1		
19h30-20h30 EDT 01h30-02h30 CET 08h30-09h30 JPN 09h30-10h30 AUS		Networking session		

Day 1 – May 10: **Rising Star Competition**
Monday May 10th, 8AM EDT/14h, CET/21h, JPN/22h AUS
Chairs: Sofie Ceberg and Brian Pogue
Each presentation will be of 10 min with an additional 2 min live for questions.

1. **Dosimetry in high-dose-rate brachytherapy with a radio-fluorogenic gel dosimeter**, K. Inoue, Y. Watanabe, T. Maeyama, A. Mochizuki, S. Mizukami, S. Hayashi, T. Terazaki, H. Muraishi, T. Gomi and T. Shimono. Kitasato University, Kanagawa; Hiroshima International University, Hiroshima; Yokohama City Minato Red Cross Hospital, Kanagawa, Japan

2. **Gel dosimetry as a tool for clinical implementation of image-guided radiotherapy.** A Elter, S Dorsch, M Marot, C Gillmann, W Johnen, A Runz, C K Spindeldreier, S Klüter, C P Karger and P Mann. DKFZ; University of Heidelberg; University Hospital Heidelberg, National Center for Radiation Research in Oncology (NCRO), Heidelberg Institute for Radiation Oncology (HIRO), HQImaging GmbH, Heidelberg, Germany

3. **Development of dosimetric procedures for ultra-high dose rate FLASH irradiation at a clinical linear accelerator**. E Konradsson, K Petersson, G Adrian, M Lempart, B Blad, S Ceberg, T Knöös, S Å J Bäck and C Ceberg. Lund University and Skåne University Hospital, Lund, Sweden; University of Oxford, Oxford, UK

4. **On the feasibility of using an optical fiber Bragg grating array for multipoint dose measurements in radiation therapy**. Marie-Anne Lebel-Cormier, Tommy Boilard, Martin Bernier and Luc Beaulieu. Département de physique, de génie physique et d'optique; Centre de recherche sur le cancer; Centre d'optique, photonique et laser (COPL); CHU de Québec - Université Laval, Université Laval, Québec, Canada

5. **Development of a practical clinical application of NIPAM kV-CBCT dosimetry**. K Pant, M Oldham, W Giles and J Adamson. Duke University and Duke University Medical Center, Durham, NC, USA

6. **The influence of internal operating temperature and supply voltages on the response of amorphous silicon electronic portal imaging devices**. J Renaud and B Muir. Metrology Research Centre, National Research Council Canada, Ottawa, Canada

7. **Dose-response stability of deformed radiochromic dosimeters during spot scanning proton therapy.** SV Jensen, L Valdetaro, PR Poulsen, P Balling, JBB Petersen and LP Muren. Danish Center for Particle Therapy, Aarhus University Hospital; Department of Physics and Astronomy, Aarhus University, Aarhus, Denmark

Networking opportunity and meet the speakers follow directly after this session: 9h30-10h30 EDT (15h30-16h30 CET; 22h30-23h30 JPN; 23h30-00h30 AUS). SpatialChat will be used for this event

Journal of Physics: Conference Series 2167 (2022) 011001 doi:10.1088/1742-6596/2167/1/011001

Day 2 – May 11 (May 12, Europe, Australia/Asia): Session 1(AUS/JPN):
Tuesday May 11th, 19h EDT / Wednesday May 12th 0h CET/ 7AM JPN/ 8AM AUS
Chairs: Kenichi Tanaka and Simon Doran
Each presentation will be of 10 min with an additional 2 min live for questions

1. Real time 4D Radiation Gel Dosimetry on the Australian MRI-Linac. Yves De Deene and Morgan Wheatley. School of Engineering, Macquarie University, North Ryde, Sydney, Australia

2. Further investigation of lung tumour peripheral doses using normoxic polymer gel dosimetry techniques. A Venning, M Mundayadan Chandroth, C Morgan and M Roberts. Port Macquarie Base Hospital, Port Macquarie, NSW, Australia

3. Influence of the components of a PVA–GTA–I radiochromic gel dosimeter on the optical dose response. Shin-ichiro Hayashi, Kaoru Ono, Keisuke Fujino, Ryosuke Kurihara. Hiroshima International University and Hiroshima Heiwa Clinic, Hiroshima, Japan

4. Measurement of spatial fluence distribution of neutrons and gamma rays using MAGAT-type gel detector doped with LiCl for BNCT at Kyoto University Reactor. Kenichi Tanaka, Yuto Ito, Yuto Murakami, Tsuyoshi Kajimoto, Shin-ichiro Hayashi, Yoshinori Sakurai, Hiroki Tanaka, Takushi Takata, Gerard Bengua and Satoru Endo. Hiroshima University, Higashi-Hiroshima; Kyoto University, Kumatori, Japan; Auckland City Hospital, Auckland, New Zealand

5. Adaptive radiotherapy dosimetry in a challenging geometry: A model gasfilled tissue expander in a helical TomoTherapy beam. T. Kairn, M. Lathouras, S. R. Sylvander, J. V. Trapp, S. B. Crowe. Royal Brisbane and Women's Hospital; Queensland University of Technology, Brisbane; Herston Biofabrication Institute, Australia

6. Investigation of target dose conformity using normoxic polymer gel dosimetry techniques: A clinical example of 12th thoracic vertebrae SBRT treatment with VMAT. A Venning, M Mundayadan Chandroth, C Morgan and M Roberts. Port Macquarie Base Hospital, Port Macquarie, NSW, Australia

7. Rapid manufacture of patient-specific, elastomeric, three-dimensional dosimeters using the FlexyDos3D dosimeter. M J Wheatley, and Y De Deene. School of Engineering, Macquarie University, North Ryde, Sydney, Australia

Networking opportunity and meet the speakers follow directly after this session: 19h00-20h30 EDT (1h-2h30 CET +1 day; 08h30-9h30 JPN, 09h30-10h30 AUS) SpatialChat will be used for this event

Day 3 – May 12 (May 13 Australia/Asia): Session 2 (EU/NA):
Wednesday May 12th, 10h EDT/16h CET/ 23h JPN/ Midnight AUS
Chairs: Sven Bäck / Mark Oldham
Each presentation will be of 10 min with an additional 2 min live for questions

1. **From conception to clinical trial: IViST, the first multi-sensor-based platform for real-time In Vivo Source Tracking in HDR brachytherapy**. H. M. Linares Rosales, Audrey Cantin, Sylviane Aubin, Sam Beddar and Luc Beaulieu. Département de physique, de génie physique et d'optique; Centre de recherche sur le cancer; CHU de Québec - Université Laval, Université Laval, Québec, Canada

2. **Optimization of solid tank design for fan-beam optical CT based 3D radiation dosimetry**. A Ogilvy, S Collins, W Hare, M Hilts, T Tuokko, R Deardon and A Jirasek. Department of Physics, University of British Columbia - Okanagan campus,; Medical Physics, BC Cancer - Kelowna, Kelowna; and Department of Mathematics & Statistics, University of Calgary, Calgary, Canada.

3. **On the use of computer vision techniques to characterize, track and correct signal variations from scintillating fibers displaced in 3D space**. Émily Cloutier, Louis Archambault and Luc Beaulieu. Département de physique, de génie physique et d'optique; Centre de recherche sur le cancer; CHU de Québec - Université Laval, Université Laval, Québec, Canada

4. **New directions for tetrazolium – gellan gum gel dosimeters**. Kalin I. Penev and Kibret Mequanint. Department of Chemical and Biochemical Engineering and Biomedical Engineering Program, University of Western Ontario and Modus Medical Devices Inc., London, Ontario, Canada

5. **Personalized pretreatment QA for intracranial stereotactic treatments using gel dosimetry and 3D printing phantom: a feasibility study**. J Colnot, S Chiavassa, G Delpon and C Huet. Institut de Radioprotection et de Sûreté Nucléaire (IRSN), Fontenay-aux-Roses, and Institut de Cancérologie de l'Ouest, Département de Physique Médicale, Site Saint-Herblain, France.

6. **End-to-End Quality Assurance of Stereotactic Radiation Therapy Using an Anthropomorphic Head Phantom**. KM Alexander, KH Dekker, T Olding and LJ Schreiner. Cancer Centre of Southeastern Ontario, Kingston Health Sciences Centre, and Department of Physics and Department of Oncology, Queen's University, Kingston, ON, Canada

7. **Validation of spine SBRT using a 3D printed anthropomorphic phantom**. KH Dekker, KM Alexander, F Ynoe de Moraes and T Olding. Cancer Centre of Southeastern Ontario, Kingston Health Sciences Centre and Department of Physics and Department of Oncology, Queen's University, Kingston, ON, Canada

Networking opportunity and meet the speakers follow directly after this session: 11h30-12h30 EDT (17h30-18h30 CET; 00h30-01h30 JPN +1 day, 01h30-02h30 AUS +1 day)
SpatialChat will be used for this event

Journal of Physics: Conference Series **2167** (2022) 011001 doi:10.1088/1742-6596/2167/1/011001

Day 4 – May 13: IC3DDose Poster Presentations
May 13, 8:00-10:30 am EDT (2:00-4:30 pm CET; 9:00 pm - 11:30 pm JPN, 10:00 pm - 00:30 am AUS)

1. Post-processing techniques using 3D Slicer for T1-weighted MRI analysis of radiochromic gel dosimeters, Hannah J. Lee, James Dolan, Stefan Pencea and Geoffrey S. Ibbott. Department of Radiation Oncology, Allegheny Health Network, Pittsburgh, PA; Department of Radiation Physics, UT MD Anderson Cancer Center, Houston, TX; Elekta, St. Charles, MO; Department of Radiation Oncology, Cooper University Health Care, Camden, NJ, USA

2. Flexible radiochromic dosimeters development for complex irradiation beams, Johanna Tran, Mathieu Agelou, Guillaume Boissonnat, Chrystèle Dehe- Pittance, Hugues Girard, Delphine Lazaro, Vesna Simic and Dominique Tromson. CEA, LIST, Laboratoire Capteurs et Architectures Electroniques, CEA Saclay,. CEA, LIST, Laboratoire de Mécanique Système et Simulation, CEA Saclay. CEA, LIST, Laboratoire Capteurs Diamant, CEA Saclay, Gif-sur-Yvette, France.

3. Three-dimensional reproducibility assessment of radiochromic ClearView™ gel in custom vessel, Kevin Jordan, Lu Xu, and Rob Barnett. Department of Medical Biophysics, Western University; Department of Oncology, Western University; London Regional Cancer Program, London Health Sciences Centre, London, ON, Canada.

4. Dose rate effects in radiochromic leuco crystal violet dosimeters, Kevin Jordan. Department of Medical Biophysics, Western University, Department of Oncology, Western University, London Regional Cancer Program, London Health Sciences Centre, London, ON, Canada.

5. Marker trajectory assessment in optical cone beam computed tomography scanner geometry, Gareth Marks Jaryd Christie and Kevin Jordan. London Regional Cancer Program, London Health Sciences Centre, Department of Medical Biophysics, Western University, Department of Oncology, Western University, London, ON, Canada

6. Optically stimulated emission of LiF:Mg,Cu,P – towards 3D optically stimulated luminescence dosimetry, Jacob S Nyemann, Mads L Jensen, Camilla L Nielsen, Karl-Emil T Bondgård, Rosana M Turtos, Brian Julsgaard, Jørgen B B Petersen, Ludvig P Muren and Peter Balling. Department of Physics and Astronomy, Interdisciplinary Nanoscience Center (iNANO), Danish Centre for Particle Therapy, Department of Medical Physics, Aarhus University and Aarhus University Hospital, Denmark

7. Synthesis and characterization of Al2O3 nanoparticles: Towards 3D optically stimulated luminescence dosimetry, Camilla L Nielsen, Martin Bondesgaard, Rosana M Turtos, Brian Julsgaard, Bo B Iversen, Ludvig P Muren and Peter Balling. Department of Physics and Astronomy, Department of Chemistry, Interdisciplinary Nanoscience Center (iNANO), Danish Center for Proton Therapy, Department of Medical Physics, Aarhus University & Aarhus University Hospital

8. Development of an automated routine for the calibration of multi-point scintillation detectors for advanced dosimetry applications, B Lessard, F Larose, F Berthiaume, S Lambert-Girard, F Therriault-Proulx and L Archambault. Département de physique, génie physique et optique, et Centre de recherche sur le cancer, Université Laval, CHU de Québec – Université Laval et CRCHU de Québec; Medscint Inc., Québec, Canada

9. Signal requirements for 3D optically stimulated luminescence dosimetry, Mads L. Jensen, Rosana M. Turtos, Jacob S. Nyemann1, Brian Julsgaard, Ludvig P. Muren and Peter Balling. Department of Physics & Astronomy, Interdisciplinary Nanoscience Center (iNANO), Danish Centre of Particle Therapy, Department of Medical Physics, Aarhus University/Aarhus University Hospital, Denmark.

10. 3D Small Field Dosimetry and Radiosurgery Plan Verification via Scintillation Imaging by Couch & Gantry Mounted Cameras, Muhammad Ramish Ashraf, Petr Bruza, Brian W. Pogue, Cedar Farwell, Benjamin B. Williams and David J. Gladstone. Thayer School of Engineering, Dartmouth College; DoseOptics; Department of Medicine, Geisel School of Medicine, Dartmouth College Hanover and Norris Cotton Cancer Center at Dartmouth-Hitchcock, Lebanon, NH, USA

11. Optical Filtering to Remove Ambient Room Light for Improved Cherenkov Imaging Dosimetry, Mahbubur Rahman, Petr Bruza, David J. Gladstone and Brian W. Pogue. Thayer School of Engineering, Dartmouth College, Hanover; Dartmouth Hitchcock Medical Center; and DoseOptics, LLC, Lebanon, NH, USA

12. Development of a novel proton CT system using a 3D scintillator detector, S Beddar, C D Darne, and D G Robertson, Department of Radiation Physics, The University of Texas MD Anderson Cancer Center, Houston, TX, USA, Department of Radiation Oncology, Mayo Clinic, Phoenix, AZ, USA

13. Development of photonic detector system for ultra-fast beam diagnostics in proton radiotherapy: the proof of concept, Viktor Iakovenko and David Jaffray. Department of Radiation Oncology, University of Toronto, ON, Canada, Department of Medical Biophysics, University of Toronto, Toronto, Ontario, Canada; Techna Institute for the Advancement of Technology for Health, Toronto, ON, Canada, M.D. Anderson Cancer Center, TX, USA

14. Utilizing Pencil Beam Scan Dynamics and a Scintillation Screen to produce 3D Dose Distribution of Proton Beams, Mahbubur Rahman, Petr Bruza, David J. Gladstone, Rongxiao Zhang and Brian W. Pogue. Thayer School of Engineering, Dartmouth College, Hanover, NH, Dartmouth Hitchcock Medical Center and DoseOptics, LLC, Lebanon, NH, USA

15. Effects of radical scavengers on nanocomposite Fricke gel for heavy ion beam irradiation, T. Maeyama, A. Mochizuki, N. Fukunishi K. L. Ishikawa, K. Fukasaku and S. Fukuda. Kitasato University/Department of Chemistry, School of Science, Kanagawa; RIKEN/Nishina Center for Accelerator-Based Science, Saitama; The University of Tokyo/Department of Nuclear Engineering and Management, Graduate School of Engineering, Tokyo; Department of Neurosurgery, Niiza-Shiki Chuo General Hospital, Saitama; RIKEN/Center for Advanced Photonics, Image Processing Research Team, National Institutes for Quantum and Radiological Science and Technology/QST Hospital, Chiba, Japan

16. A 3D printed eyeball phantom for Sr-90 dosimetry measurements, S K Maxwell1, S B Crowe, E M Simpson-Page, N Cassim and T Kairn. Royal Brisbane and Women's Hospital, and Queensland University of Technology, Brisbane QLD; Herston Biofabrication Institute, Herston, QLD, Australia

17. Monte Carlo simulations of a non-invasive positron detector to measure the arterial input function for dynamic PET, Liam Carroll and Shirin A. Enger. Medical Physics Units and Department of Oncology, McGill University, Research Institute of the McGill University Health Centre and Lady Davis Research Institute for Medical Research, Jewish General Hospital, Montreal, Quebec, Canada

Peer Review Statement

All papers published in this volume have been reviewed through processes administered by the Editors. Reviews were conducted by expert referees to the professional and scientific standards expected of a proceedings journal published by IOP Publishing.

- **Type of peer review:** Double Anonymous
- **Conference submission management system:** Morressier
- **Number of submissions received:** 80
- **Number of submissions sent for review:** 80
- **Number of submissions accepted:** 34
- **Acceptance Rate (Submissions Accepted / Submissions Received × 100):**
- **Average number of reviews per paper:** 2
- **Total number of reviewers involved:** 16
- **Contact person for queries:**
  Name: Clive Baldock
  Email: c.baldock@westernsydney.edu.au
  Affiliation: Western Sydney University

Post-processing techniques using 3D Slicer for T1-weighted MRI analysis of radiochromic gel dosimeters

Hannah J. Lee[1,2], James Dolan[3], Stefan Pencea[3], Geoffrey S. Ibbott[4,2]
[1]Department of Radiation Oncology, Allegheny Health Network, Pittsburgh, PA, USA
[2]Department of Radiation Physics, UT MD Anderson Cancer Center, Houston, TX, USA
[3]Elekta, St. Charles, MO, USA
[4]Department of Radiation Oncology, Cooper University Health Care, Camden, NJ, USA

E-mail: hannah.lee@ahn.org

Abstract. Radiochromic FXG gel dosimeters were investigated for end-to-end validation of TG-119 plans delivered in the pre-clinical Elekta 1.5 T-7 MV FFF MR-Linac. Due to the noise levels in the T1-weighted MR images, this preliminary study investigated post-processing techniques available in 3D Slicer and their impact on gamma passing rates. Binomial blur and discrete Gaussian image filters were identified as the most impactful on increasing gamma passing rates for T1-weighted MR images; however, any application of image filters should be implemented with caution on the interpretation of the results.

1. Introduction

MR image-guided radiation therapy (MR-IGRT) systems require a modified clinical workflow to integrate MR images for daily adaptive re-planning. Various methods for confirmation of accurate treatment plan delivery have been performed using quasi-3D arrays, 2D arrays, and 3D dosimeters [1-12]. Ideally, a heterogeneous, anthropomorphic MR-visible and radiation-sensitive phantom would be used for a complete end-to-end clinical workflow assessment. Although commissioning phantoms using CT and MR-visible materials are in development, these depend on point and planar measurements using thermoluminescent dosimeters and films, which can miss dose information in three dimensions [13-15].

To meet this quality assurance (QA) need, end-to-end dosimetry with homogeneous and heterogeneous 3D dosimeters has been investigated using optical and MRI read-out techniques [5, 7-9, 12]. Prior to MR-IGRT systems, it was not possible to pre-scan, irradiate, and post-scan 3D dosimeters all in the same position. Several MR-IGRT systems are now in clinical use world-side and others are still in development, including the 1.5 MRI – 7 MV FFF linear accelerator (linac) (Unity MR-Linac, Elekta AB, Stockholm, Sweden), a prototype system combining a 1.0 T MRI with a 6 MV linac (Ingram Institute, Sydney, Australia), the Aurora RT 0.5 T magnet with 6 MV linac (MagnetTx, Alberta, Canada), and the MRIdian 0.35 T magnet with 6 MV linac (Viewray, Inc., Oakwood Village, Ohio, USA). With the MR-Linac and other MR-IGRT systems, we now have the capability for delivery of a treatment plan to a 3D dosimeter in the exact position that it was planned for with the possibility for real-time onboard MR imaging during irradiation and immediate MR imaging post-irradiation [8, 16]. Comparisons between treatment plan dose and measured 3D dosimeter dose with gamma passing rates greater than 90% for 3%/3mm distance to agreement (DTA) have been reported for non-MR visible plastic PRESAGE (optical read-out) and polymer gel (spin-spin relaxation rate R_2 (=1/T_2) MRI read-out) dosimeters. In this study, we investigate post-processing techniques to modify the MR image quality for radiochromic gel (spin-lattice relaxation rate R_1 (=1/T_1) MRI read-out) dosimeters that were irradiated in a pre-clinical Elekta 1.5 T-7 MV FFF MR-Linac and assess the impact on gamma pass rates.

2. Materials and Methods

2.1. FXG gel fabrication and MR imaging

Radiochromic gel dosimeters consist of a radiochromic reporter compound, a free radical source and solvent, an iron source, and a gelling agent. The Fricke xylenol orange gel (FXG) formulation was as follows (all chemicals sourced from Sigma-Aldrich): 0.05 mM xylenol orange disodium salt, ~96% wt% deionized water, 1 mM ferrous ammonium sulfate, 4 wt% 300 bloom gelatin, and 50 mM sulfuric acid. The mixture was poured into two liter-sized uniform phantoms then stored at 4°C and acclimated to room temperature prior to irradiation.

Pre-irradiation and post-irradiation images were acquired for each phantom using T1-weighted contrast enhanced (CE) MRI sequences (3D fast field echo (FFE), TR/TE = 11/4.6 ms, 400 x 400 x 300 mm^3 field of view, reconstructed voxels of 0.83 x 0.83 x 1.00 mm^3). T1-weighted CE and no CE are Philips-specific terminology for their 3D FFE sequences and do not indicate that an MR contrast agent was injected. The T1-weighted CE sequences spoiled the transverse magnetization using pulse phase cycling of the radiofrequency excitation pulses. Therefore, the no CE images included a mixed signal of the free induction decay (FID) and spin echo, and the T1-weighted CE images only contained the FID signal.

2.2. Treatment planning and plan delivery

Reference step-and-shoot IMRT plans based on the TG-119 IMRT commissioning tests were created in a Monaco treatment planning system (TPS) (research version 5.19.02) using CT images acquired of the homogeneous cylindrical phantoms [5, 17, 18]. Monaco TPS for the Elekta MR-Linac used Monte Carlo dose calculations in a graphical processing unit environment that incorporated a 1.5 T B_0 field. Monaco plan doses were calculated with 1-3% statistical uncertainty per control point and 0.2-0.3 cm grid spacing. At the time of plan delivery, complete cross-validation of the Monaco TPS had not yet been done, and greater uncertainties in dose were expected (conventional TPS calculations include up to 3% dose uncertainties in clinical use due to different factors). The Elekta MR-Linac was also in its pre-clinical phase at the time of plan delivery (prior to final system upgrades to the CE marked and 510(k) cleared Unity system currently in clinical use). Most notably, rigorous MLC calibration and MR to MV isocenter registration had not yet been completed at the time of plan delivery, affecting the accuracy and delivery of adapted plans based on daily MR images. A workflow comparable to the current clinical adapt-to-position (ATP) was used for treatment delivery. Daily T1-weighted no CE MR images (same MRI sequence parameters as for T1-weighted CE described above) were acquired for fusion with the CT to create the daily adapted plan based on the phantom position in the pre-clinical MR-Linac. MR images acquired with T1-weighted no CE were found to have superior visibility of edges and reduced noise compared to T2-weighted and T1-weighted CE images; however T1-weighted no CE could not be used for post-processing of dose for the FXG formulation [5]. These daily adapted plans were then delivered in the pre-clinical MR-Linac and post irradiation MR images were acquired of the gels in the same position as described above.

2.3. Gamma analysis and post-processing with 3D Slicer

After acquisition of the pre-irradiation and post-irradiation MR images, the relative volumetric dose distributions quantified in the gels were compared to calculated doses from the Monaco TPS using 3D Slicer (version 4.6.2) [19, 20]. 3D Slicer was an open source software platform that could be used for a wide variety of applications and has been validated for TPS plan comparisons using gamma analysis

[21–23]. The 3D Slicer workflow has been described previously and is summarized below: using the SlicerRT extension, import and load DICOM files of the TPS dose, structures, pre-irradiation MRI, and post-irradiation MRI); subtract the MRI prescan from the MRI postscan; transform the subtracted MRI volume to match TPS plan volume using the adapted plan fusion coordinates; post-process as needed and scale to relative dose; assess 3D gamma analysis using the Dose Comparison module [5].

A gamma criterion of 7%/4mm DTA was used by IROC for IMRT head and neck plans for 2D gamma analysis of film [24]. While 3D dosimetry studies have used other gamma criteria for 3D gamma analysis along with other metrics for plan delivery assessment, this initial study utilized 7%/4mm along with stricter gamma constraints (5%/3mm and 3%/3mm) as a comparison with passing cut-offs of 80% [24–27]. Up to approximately 5% differences in gamma passing rates were found depending on small changes to scaling for relative doses. A 10% maximum dose threshold (TH) was applied for all gamma analysis, regardless of the gamma criteria. Due to the standard deviation as a result of noise in MR images (up to ~10% of the mean MR signal intensity), it should be noted that there is a risk of voxels passing with 3D gamma analysis using loose criteria such as 7%/4mm.

Image filters that could potentially improve the MR image quality were assessed. The 3D slicer post-processing tools used for this study were the following image filters: median, mean, binomial blur, box mean, curvature flow, discrete Gaussian, grayscale grind peak, Laplacian sharpening, min max curvature flow, rank, and smoothing recursive Gaussian. These image filters included a variety of techniques, including direct (such as median) and recursive (rank), linear (mean) and nonlinear (median), and retention (median) and sharpening (Laplacian sharpening) of boundaries. After the optimal image filters were identified, a combination using the histogram matching image filter was also investigated. The use of any image filter should be implemented with caution to modify the raw data set, which could also perturb the data in an unexpected manner and may result in hiding issues with the plan delivery.

3. Results and Discussions

3.1. Post-processing filters

The post-processing image filters listed above were first assessed for the TG-119 AP PA plan. A variety of parameters and number of iterations were applied for each image filter. Across all applied image filters, the gamma passing rates were 98.55 ± 0.32% for 7%/4mm criteria; 90.83 ± 2.72% for 5%/3mm; and 79.39 ± 5.24% for 3%/3mm. The image filters resulting in the highest 3%/3mm gamma passing rates were binomial blur image filter (Repetitions 2), discrete Gaussian image filter (Variance 1), and smoothing recursive Gaussian image filter (Sigma 1). The image filters resulting in the highest 7%/4mm gamma passing rates were curvature flow image filter (# Iterations 3), min max curvature flow image filter (# Iterations 5), curvature flow image filter (# Iterations 5), and binomial blur image filter (Repetitions 1). Due to potentially misleading higher passing rates for 7%/4mm, the binomial blur (used nearest neighbor averaging) and discrete Gaussian (used separable convolution with discrete Gaussian kernels) image filters were further assessed using a combination.

Applications of the binomial blur image filter and discrete Gaussian image filter resulted in comparable gamma passing rates with lower NSA (number of signal averages, also known as number of excitations NEX) to the application of simpler image filters, such as the median image filter, with higher NSA. Increasing NSA increased the acquisition time linearly (1:54 minutes, 5:42 minutes, and 9:29 minutes for NSA = 1, 3, and 5, respectively). The use of greater NSA may still be required for analysis of MR images acquired at very low field strengths to improve SNR [12]. The combination of binomial blur and discrete Gaussian image filters did not guarantee higher gamma passing rates for higher NSA.

3.2. TG-119 Plan Analysis

Using the binomial blur, discrete Gaussian, and a combination using the histogram matching image filter, the gamma passing rates for all TG-119 plans are listed in Table 1 with all images acquired using T1-weighted CE and NSA = 1 sequences.

Table 1. 3D gamma passing rates for TG-119 plans (10% TH). >80% passing rates in green.

TG-119 Plan	Post-processing	7%/4mm	5%/3mm	3%/3mm
AP PA	Binomial Blur Image Filter	98.25%	92.34%	83.71%
	Discrete Gaussian Image Filter	98.03%	92.00%	83.69%
	Histogram Matching Image Filter	98.15%	92.27%	84.02%
	Median Image Filter	98.68%	92.24%	82.01%
MultiTarget	Binomial Blur Image Filter	95.96%	53.85%	34.43%
	Discrete Gaussian Image Filter	95.78%	55.52%	35.54%
	Histogram Matching Image Filter	95.73%	54.20%	34.80%
	Median Image Filter	96.07%	48.50%	30.62%
Prostate	Binomial Blur Image Filter	95.69%	46.93%	28.49%
	Discrete Gaussian Image Filter	95.77%	47.94%	28.79%
	Histogram Matching Image Filter	95.97%	48.48%	29.53%
	Median Image Filter	96.12%	47.16%	29.12%
Head/Neck	Binomial Blur Image Filter	94.05%	48.53%	30.07%
	Discrete Gaussian Image Filter	93.88%	49.90%	30.85%
	Histogram Matching Image Filter	94.33%	50.04%	31.07%
	Median Image Filter	94.28%	44.82%	27.56%
C-Shape	Binomial Blur Image Filter	94.64%	51.04%	31.86%
	Discrete Gaussian Image Filter	94.43%	51.90%	32.49%
	Histogram Matching Image Filter	95.01%	52.60%	32.94%
	Median Image Filter	94.79%	47.94%	29.99%

The homogeneous large FXG dosimeters used for TG-119 IMRT plan evaluation passed all 7%/4mm gamma criteria. However, the other more complex plans had gamma passing rates that dropped more steeply for 5%/3mm and 3%/3mm gamma criteria, which could be attributed to the longer plan delivery times for the complex step-and-shoot IMRT plans as well as the limitations to this pre-clinical system.

4. Conclusions

As shown in this study and in the literature, the post processing of measured signal can significantly affect the gamma passing rate, and thus needs to be specified in the QA protocol. The results of this study demonstrated that optimizing post-processing techniques can improve gamma passing rates with the binomial blur and discrete Gaussian image filters identified as the most impactful for T1-weighted CE MR images. This study demonstrated using an MR-visible 3D dosimeter to eliminate registration errors between pre-irradiation imaging, plan delivery, and post-irradiation imaging of the dosimeter.

Future work in a fully clinically commissioned MR-Linac system should investigate further improving the combination of MR image acquisition sequence parameters with considerations for diffusion and post-processing to improve gamma passing rates for tighter gamma criteria as well as investigating additional metrics to gamma passing rate for analysis of the plan delivery quality.

5. References

[1] Chen X, Paulson ES, Ahunbay E *et al* 2019 *J. Appl. Clin. Med. Phys.* **20** 28-38
[2] de Vries JHW, Seravalli E, Houweling AC *et al* 2018 *Phys. Med. Biol.* **63** 02NT02
[3] Ellefson ST, Culberson WS, Bednarz BP 2017 *J. Appl. Clin. Med. Phys.* **18** 161-71
[4] Houweling AC, de Vries JHW, Wolthaus J *et al* 2016 *Phys. Med. Biol.* **61** N80-9
[5] Lee H 2017 *Volumetric, Magnetic Resonance-Visible, and Radiation-Sensitive Detectors for Magnetic Resonance Image-Guided Radiation Therapy*. The University of Texas, MD Anderson Cancer Center, UTHealth Graduate School of Biomedical Sciences. http://digitalcommons.library.tmc.edu/utgsbs_dissertations/803
[6] Li HH, Rodriguez VL, Green OL *et al* 2015 *Int. J. Radiat. Oncol.* 91 65-72

[7] Mein S, Rankine L, Adamovics J *et al* 2017 *Med. Phys.* **44** 6018-28
[8] Pappas E, Kalaitzakis G, Boursianis T *et al* 2019 *Phys. Med. Biol.* **64** 225009
[9] Roed Y 2017 *Polymer Gel Dosimetry in Magnetic Resonance-Image Guided Radiation Therapy.* University of Houston. https://uh-ir.tdl.org/uh-ir/handle/10657/2880
[10] Smit K 2015. *Dosimetry for the MR-linac.* https://dspace.library.uu.nl/bitstream/1874/318363/1/Smit.pdf
[11] Smit K, Kok JGM, Lagendijk JJW *et al* 2014 *Phys. Med. Biol.* **59** 1845-55
[12] Zhang L, Hu Y, Du D *et al* 2015 *J. Med. Phys.* **40** 176-80
[13] Steinmann A, Alvarez P, Lee H *et al* 2020 *Med. Phys.* **47** 604-13
[14] Steinmann A, Alvarez P, Lee H *et al* 2019 *Med. Phys.* **46** 5124-33
[15] Steinmann A, Stafford RJ, Sawakuchi G *et al* 2018 *Med. Phys.* **45** 773-82
[16] Lee HJ, Alqathami M, Kadbi M *et al* 2016 *Int. J. Radiat. Oncol.* **96** E631
[17] Ezzell GA, Burmeister JW, Dogan N *et al* 2009 *Med. Phys.* **36** 5359-73
[18] TG-119 IMRT 2009 Commissioning Tests Instructions for Planning, Measurement, and Analysis. *Med. Phys.* http://www.aapm.org/pubs/tg119/TG119_Instructions_102109.pdf
[19] 3D Slicer https://www.slicer.org/ (accessed 22 July 2017).
[20] Fedorov A, Beichel R, Kalpathy-Cramer J *et al* 2021 *Magn. Reson. Imaging* **30** 1323-41
[21] Pinter C, Lasso A, Wang A *et al* 2012 *Med. Phys.* **39** 6332-8
[22] Choi GW 2016. *Measurement of the Electron Return Effect Using PRESAGE Dosimeter.* The University of Texas. http://digitalcommons.library.tmc.edu/utgsbs_dissertations/690
[23] Alexander KM, Pinter C, Fichtinger G, et al. Streamlined open-source gel dosimetry analysis in 3D slicer. *Biomed Phys Eng Express* 2018; 4: 45041.
[24] Ibbott GS, Followill DS, Molineu HA *et al* 2008 *Int. J. Radiat. Oncol.* **71** S71-5
[25] Low DA, Harms WB, Mutic S *et al* 1998 *Med. Phys.* **25** 656-61
[26] Hussein M, Clark CH, Nisbet A 2017 *Phys. Medica* **36** 1-11
[27] Pulliam KB, Huang JY, Howell RM 2014 *Med. Phys.* **41** 21710

Flexible radiochromic dosimeters development for complex irradiation beams

Johanna Tran[1], Mathieu Agelou[2], Marie-Noëlle Amiot[3], Guillaume Boissonnat[2], Chrystèle Dehe-Pittance[1], Hugues Girard[4], Delphine Lazaro[2], Vesna Simic[1] and Dominique Tromson[1]

[1]CEA, LIST, Laboratoire Capteurs et Architectures Electroniques, CEA Saclay, 91191 Gif-sur-Yvette, France

[2]CEA, LIST, Laboratoire de Modélisation des Systèmes et Simulation, CEA Saclay, 91191 Gif-sur-Yvette, France

[3]CEA, LIST, Laboratoire de Mesure de la Dose, CEA Saclay, 91191 Gif-sur-Yvette, France

[4]CEA, NIMBE, Laboratoire Edifices Nanométriques, CEA Saclay, 91191 Gif-sur-Yvette, France

Email: johanna.tran@cea.fr, dominique.tromson@cea.fr

Abstract. In vivo dosimetry for patient radioprotection is a real challenge concerning new radiotherapy treatments. In this study we aim to develop a dosimeter based on tissue equivalent material, flexible and adaptable to the patient morphology in order to perform in vivo dosimetry for complex irradiation beams. Here, we report the evaluation in standard beams of highly flexible dosimeter composed of a silicone elastomer and containing leucomalachite green as radiochromic dye and give a first estimation of LMG-micelle hydrogel response. All results are compared to a commercially available PRESAGE® dosimeter.

1. Introduction

Despite the emergence of several technological innovations in the field of external radiotherapy, patient radioprotection is still a real challenge, particularly in complex beams. Indeed, radiotherapy has gained more and more complexity making failure possible at different steps of the treatment. Dosimetry and mostly *in vivo* dosimetry is taking a central and crucial role in quality insurance.

The use of radiation sensitive matrix for dosimetry purposes was proposed as early as the 1950s [1]. Since then, new formulations [2-4] have been developed such as PRESAGE® [5-7] which contains leucomalachite green (LMG) as leucodye in a polyurethane matrix. Despite some advantages PRESAGE® presents the main drawback for our application of being a hard matrix.

Research on dosimeter has focused on gaining more and more sensitivity as well as lower the dose detected. A way to enhance sensitivity, is to incorporate high-Z compounds in the dosimeter matrix to increase the dose response of the material.

Here we present the results of a highly flexible LMG-based dosimeter contained in a silicone elastomer matrix and of a polymer gel made of Pluronic®, these could be mould to exactly fit the patient to obtain accurate *in vivo* dosimeter. Dosimetric parameters were evaluated as dose, dose rate and energy dependency in standard beam. An evaluation of the dosimeter in flattening filter-free (FFF) mode was performed. Our results were compared to a new version of commercially available PRESAGE® dosimeter.

2. Experimental Part

2.1. Materials
Reagents were used as received without further purification. SYLGARD® 184 Silicone Elastomer Kit was purchased from Dow Corning. Leucomalachite Green (LMG) (4,4′-Benzylidenebis(N,N-dimethylaniline), Pluronic® F127, Tween 80 (polysorbate 80) were purchased from Sigma Aldrich (Merck). Chloroform was purchased from VWR. PRESAGE® dosimeter were purchased from John Adamovics. PRESAGE® formulation: 2-methyl-LMG-DEA, trimethylbenzene, carbon tetrabromide and polyurethane

2.2. Fabrication

2.2.1. LMG elastomer. SYLGARD® 184 Silicone Elastomer Kit was used. The polymer base and the curing agent were used in a 10:1 weight ratio. The dosimeter contained 0.24 wt% of LMG and 1.9 wt% of chloroform. A planetary centrifugal mixer (Thinky, ARE-250) was used to mix all the components together.

A batch of 14 cuvettes were prepared as follows. In an adapted Thinky mixer container, 50 g of SYLGARD® 184 Silicone Elastomer Base and 5 g of SYLGARD® 184 Silicone Elastomer Curing Agent were poured. The silicone mixture was mixed for 1 min at 900 rpm using the planetary mixer. Meanwhile, in a small vial 137.5 mg of LMG was dissolved in 370 μL (548 mg) of chloroform. The organic solution was poured in the matrix mix and the container was mixed for 6 min at 900 rpm. Finally, 370 μL (548 mg) of chloroform was put in the container. The mixture was stirred for an additional 1 min at 900 rpm. The mixture was poured into disposable PMMA cuvettes (1x1x4.5 cm^3), closed with caps and let allowed to cure in the dark at room temperature for 48 hours.

2.2.2. LMG micelles in Pluronic. First, the micelles were synthesized by the nanoprecipitation technique [8] and Tween 80 was used as surfactant. Then, a solution of Pluronic® F127 was poured onto the micelles to form the gel. The gel contained 0.22 wt% LMG and 1.47 wt% of chloroform. A viscous solution of Pluronic® F127 30 wt% was prepared prior to the preparation of the radiochromic hydrogel and kept in the fridge. Magnetic stirring was used to prepare the gel.

A batch of 12 cuvettes were prepared as follows. In a beaker with a magnetic bar, 1.5 g of viscous Tween 80 was solubilized in 25 g of distilled water. Meanwhile, in a small vial 140 mg of LMG was dissolved in 625 μL (925 mg) of chloroform. This organic solution was poured drop by drop onto the Tween 80 solution under magnetic stirring, leading to the formation of micelles that have a size of about 40 nm (size given by dynamic light scattering (DLS)). Working in an ice bath to avoid quick gel formation, 35 g of Pluronic® F127 30 wt% solution was added to the micelles solution under magnetic stirring. The mixture was then poured into disposable PMMA cuvettes (1x1x4.5 cm^3), closed with caps and allowed to reach room temperature for gelification.

2.3. Irradiation
Irradiations were performed on a Versa-HD System (ELEKTA) at DOSEO Platform (CEA Saclay). The cuvettes were irradiated in a 10 x 10 cm^2 field whether with an energy of 6 MV or 18 MV. 0.7, 1.3, 2.6 and 5.3.Gy.min^{-1} using the flattening filter and 13.7 Gy.min^{-1} in the flattening filter-free (FFF) mode The cuvettes were placed between two PMMA sheets in order to have a Source-Surface Distance (SSD) at 97.5 cm and a Source-Detector Distance at 100 cm.

2.4. Dose read-out and data analysis
The absorbance of the cuvettes was measured with a spectrophotometer (Cintra 4040, GBC Scientific Equipment Pty Ltd) between 350 and 800 nm. The spectra before irradiation were measured maximum

Journal of Physics: Conference Series **2167** (2022) 012002 doi:10.1088/1742-6596/2167/1/012002

within an hour before irradiation and after irradiation within a half-hour. Data were analysed using GraphPad Prism, all spectra were baseline corrected using the average between 750 and 800 nm.The optical change caused by the irradiation was determined, for each formulation, by measuring pre- and post-irradiation absorbance at the maximum of absorption (*i.e.* 627 nm for LMG-based material and 635 nm for PRESAGE®).

3. Results & Discussion

LMG elastomer, LMG micelles and PRESAGE® were irradiated using a Versa-HD System at 6 MV with a dose-rate of 5.3 $Gy.min^{-1}$. The materials' dose response was taken as the slope of the curve fit of the difference of absorbance after and before irradiation (Δabs) against the dose (Figure 1). The dose response of LMG elastomer is 2.2 x 10^{-2} $Gy^{-1}.cm^{-1}$ which is slightly higher than the value we obtained for PRESAGE® (1.6 x 10^{-2} $Gy^{-1}.cm^{-1}$). Yet, this value is 4-fold higher than the one reported for a previous formulation of PRESAGE® [9]. Note that the dose response of LMG elastomer could change depending of the LMG batch. The lowest detectable dose was 0.3 Gy for LMG elastomer whereas 0.6 Gy for PRESAGE®.

LMG micelles present the lowest dose response of the study (5.7 x 10^{-3} $Gy^{-1}.cm^{-1}$) even though the amount of LMG and chloroform in the hydrogel is quite comparable to the quantities in the LMG elastomer. While, dose response is low in comparison to LMG elastomer, LMG micelles hydrogel is transparent and water-based, making it easier to incorporate aqueous gold or silver-based metallic nanoparticles in order to increase the sensitivity. Indeed, incorporation of high-Z compounds in the matrix could increase the number of secondary electrons generated by physical interactions between photons and the material, thus increase the dose response.

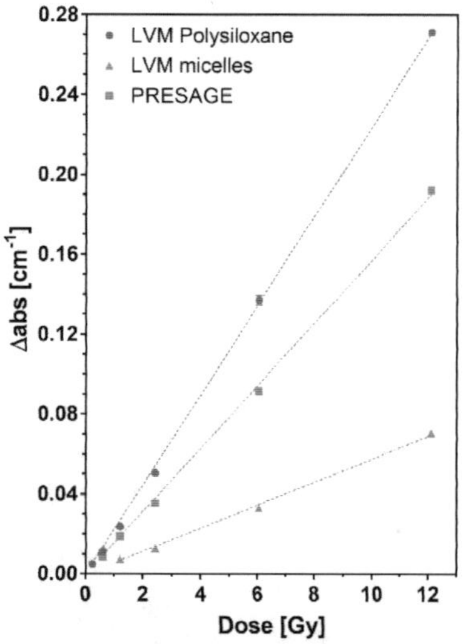

Figure 1. Difference of absorbance after and before irradiation as a function of dose for three different dosimeters. Materials were irradiated at 6 MV with a dose-rate of 5.3 $Gy.min^{-1}$. LMG elastomer (circles), LMG micelles (triangles) and PRESAGE® (squares).

LMG elastomer and PRESAGE® were also evaluated at another energy, *i.e.* 18 MV (Figure 2). A small difference in dose response was observed for the LMG elastomer samples that were irradiated with a 6 MV photon beam than those irradiated with an 18 MV photon beam, whereas a bigger difference was observed for PRESAGE® samples. Small energy dependency is definitively an advantage towards LMG elastomer.

Both materials were also tested at different dose-rate at an energy of 6 MV and an irradiation dose of 1.2 Gy (Figure 3). For PRESAGE® samples, it is difficult to discern a trend, indeed the experimental points are dispersed quite hectically (data point for the dose-rate 2.6 $Gy.min^{-1}$ is hidden by the LMG elastomer point). On the contrary, LMG elastomer showed a decrease of sensitivity with the increase of dose-rate. With the flattening filter (*i.e.* for the following dose rate: 0.7, 1.3, 2.6 and 5.3.$Gy.min^{-1}$) a decrease of about 10% was observed relative to 0.7 $Gy.min^{-1}$, in flattening filter-free (FFF) mode (*i.e.* 13.7 $Gy.min^{-1}$) a decrease of about 25% was observed still relative to 0.7 $Gy.min^{-1}$. The decrease is relatively small and maybe by increasing the number of samples, we would see no dependency with the dose-rate.

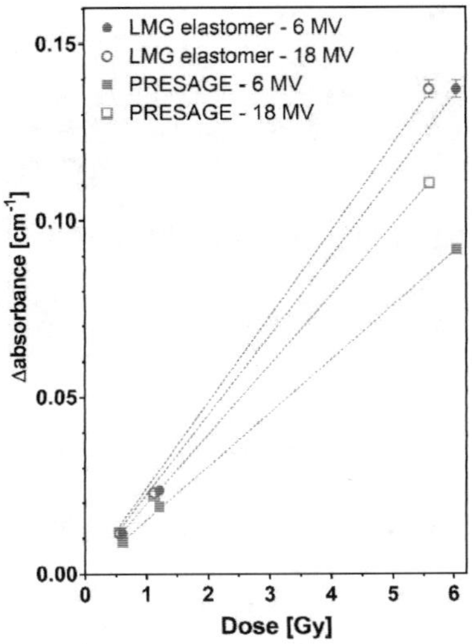

Figure 2. Difference of absorbance after and before irradiation against dose. Materials were irradiated whether at 6 MV (full symbols) or at 18 MV (unfilled symbols), with a dose rate of 5.3 Gy.min^{-1}. LMG elastomer (circles) and PRESAGE® (squares).

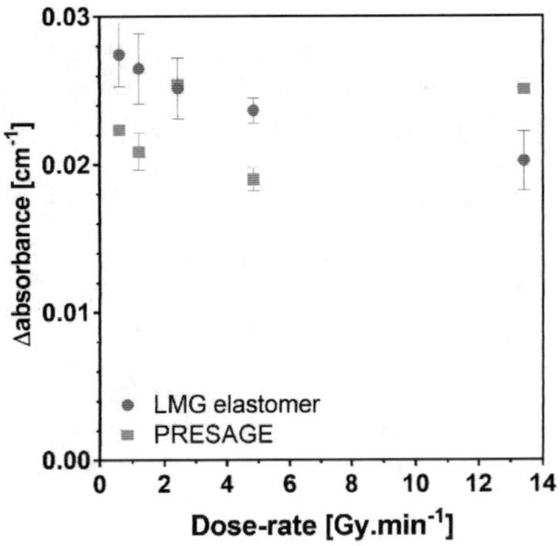

Figure 3. Difference of absorbance after and before irradiation against dose-rate. Materials were irradiated at 6 MV with a dose of 1.2 Gy. LMG elastomer (circles) and PRESAGE® (squares).

4. Conclusion

This silicone-based elastomer is highly potent for *in vivo* dosimetry. The aim of the project is to have a flexible dosimeter able to fit patient body and the highest sensitivity as possible. It presents a comparable sensitivity to PRESAGE® but above all, no energy dependence and relatively low dose-rate dependency. LMG micelle dosimeter was evaluated in this study due to the possibility to incorporate quite easily nanoparticles, relatively low dose sensitivity was found for this material. Further investigations will be carried out related to nanoparticles preparation in solvent to directly incorporate in LMG elastomer in order to combine good elastomer properties and increase in sensitivity.

5. Acknowledgment

The authors would like to thank Christel Stien (CEA, LIST, Laboratoire Métrologie de la Dose) for lending her spectrophotometer. This work was realised through the FaNTOM project with the financial support provided by ITMO Cancer.

6. References

[1] Baldock C *et al* 2010 *Phys. Med. Biol.* **5** R1-R63
[2] Babic S, Battista J, Jordan K 2008 *Phys. Med. Biol.* **53** 1637-50
[3] Kozicki M, Jaszczak M, Maras P *et al* 2017 *Phys. Med. Biol.* **62** 986-1008
[4] Baldock C 2017 *J. Phys.: Conf. Ser.* **777** 012029
[5] Adamovics J, Maryanski M 2006 *Radiat. Prot. Dosim.* **120** 107-12
[6] Alqathami M *et al* 2013 *Radiat. Phys. Chem.* **85** 204-9
[7] Khezerloo D *et al* 2017 *Radiat. Phys. Chem.* **141** 88-97
[8] Fessi H, Puisieux F, Devissaguet J P *et al* 1989 *Int. J. Pharm.* **55** R1-4
[9] Alqathami M *et al* 2012 *Radiat. Phys. Chem.* **81** 1688-95

Development of dosimetric procedures for experimental ultra-high dose rate irradiation at a clinical linear accelerator

E Konradsson[1], K Petersson[2,3], G Adrian[4], M Lempart[2], B Blad[2], S Ceberg[1], T Knöös[1,2], S Å J Bäck[1,2] and C Ceberg[1]
[1]Medical Radiation Physics, Department of Clinical Sciences, Lund University, Lund, Sweden
[2]Radiation Physics, Department of Hematology, Oncology and Radiation Physics, Skåne University Hospital, Lund, Sweden
[3]Oxford Institute for Radiation Oncology, Department of Oncology, University of Oxford, Oxford, UK
[4]Division of Oncology and Pathology, Department of Clinical Sciences, Lund University, Lund, Sweden

E-mail: elise.konradsson@med.lu.se

Abstract. As radiotherapy using ultra-high dose rates has gained new interest, the dosimetric challenges arising at these conditions needs to be addressed. Ionization chambers suffer from a large decrease in ion collection efficiency due to ion recombination, making on-line dosimetry difficult. In this work we present experimental setups and dosimetric procedures for FLASH irradiation of cells, zebrafish embryos and small animals using a 10 MeV electron beam at a modified clinical linear accelerator, and describe the dosimetric steps required to initiate clinical trials. The dosimetric equipment used for our pre-clinical experiments consisted of radiochromic film, thermoluminescent dosimeters, a Farmer-type ionization chamber and phantom material mimicking the experimental setup for irradiation. In preparation for small animal irradiation, dose profiles and depth dose curves were measured for all collimator sizes. The average dose rates were ≥620 Gy/s, ≥640 Gy/s and ≥400 Gy/s for cells, zebrafish embryos and small animals, respectively.

1. Introduction

Recent discoveries have drawn new attention to radiotherapy delivered with ultra-high dose rates (>40 Gy/s), i.e. FLASH radiotherapy. In several pre-clinical *in vivo* studies, irradiation with ultra-high dose rates has shown to reduce normal tissue toxicity compared to conventional radiotherapy [1-7]. Yet, further radiobiological research remains to elucidate the underlying mechanism of this effect. Our research group has previously modified a clinical linear accelerator to enable delivery of ultra-high dose rate electrons [8], and as a step towards clinical trials with this novel technique we have conducted several radiobiological experiments, such as presented by Adrian *et al.* [9]. However, dosimetry at these extreme conditions is complicated and there are some associated technical challenges that must be addressed. As dose rate increases from conventional (a few Gy/min) to ultra-high (>10's of Gy/s), the ionization chambers commonly used for on-line dose measurements are affected by ion recombination, causing a drop in ion collection efficiency [10,11]. Thus, there is a need for novel dosimetric procedures that enables accurate delivery of the desired dose in FLASH irradiations.

In this work we present experimental setups and dosimetric procedures for FLASH irradiation of cells, zebrafish embryos and small animals, as well as a proposed dosimetric procedure for upcoming clinical trials on veterinary and subsequently human patients, using a modified clinical linear accelerator.

2. Materials and methods

2.1. Irradiation source

Our irradiation source is a clinical linear accelerator (Elekta AB, Stockholm, Sweden) modified for FLASH irradiation with a 10 MeV electron beam, previously described by Lempart *et al.* [8]. To achieve maximal radiation output, the accelerator was operated without the primary and secondary scattering foils. The radiation was delivered in short (3.5 μs) pulses with a pulse repetition frequency of 200 Hz.

2.2. Setup, dosimetric procedure and beam characterization

The dosimetric equipment consisted of dose rate independent [12,13] radiochromic film (GafChromic EBT^3 and EBT-XD) and thermoluminescent dosimeters (TLD), as well as a Farmer-type ionization chamber. Film and TLD were calibrated against an ionization chamber traceable to a standard laboratory for a dose range of 1-30 Gy (1-40 Gy for EBT-XD film). The agreement between dose measured with EBT^3 and TLD was determined by simultaneous measurements at 20 mm depth in a solid water phantom. The Farmer-type ionization chamber was used as a relative dosimeter at low dose rates (<0.1 Gy/s), where ion recombination is not an issue.

2.2.1. Irradiation of cells and zebrafish embryos. For cell irradiations, the gantry angle was at 180° and T12.5 cell flasks (Thermo-Fisher Scientific™, Waltham, MA) were placed one-by-one on top of the cross-hair foil in the centre of the beam. The collimators were set to their outer limits, i.e. creating a field size of 40x40 cm^2 at isocentre distance. The beam profile at the position of the cross-hair foil was measured using EBT^3 film at dose maximum (15 mm depth) in a polystyrene phantom, and the full width at half maximum (FWHM) and penumbra width were determined. Before and after cell irradiations, EBT^3 film measurements were carried out by placing the film at the bottom inside a cell flask, under a 2 mm sheet of polystyrene mimicking cell media (Figure 1). To ensure accurate dose delivery during the cell irradiations, the Farmer-type ionization chamber was positioned in the ceiling where the dose rate was low enough to ensure dose rate independence (although high enough to be measurable), making it useful as a relative dosimeter to correlate to the film measurements. A similar setup was established for irradiation of zebrafish embryos, but instead of using flasks, embryos were collected in 1.5 ml Eppendorf tubes that were fitted into a 3D printed plastic phantom (Figure 2). Dosimetric measurements were carried out before and after irradiation of the embryos by 1) fitting a 3D printed insert (in the shape of an Eppendorf tube), with TLD slots, into the phantom, and 2) using a similar 3D printed phantom constructed for film measurements (Figure 3).

2.2.2. Small animal irradiations. An experimental setup for small animal irradiations was established using an electron applicator with a source-to-applicator-end distance of 650 mm. Cerrobend blocks of various sizes were created for field collimation, and for each size; dose profiles, output factors (at 20 mm depth, i.e. at depth of dose maximum) and depth dose curves (0-42 mm depth) were measured in a solid water phantom using EBT-XD film. FWHMs and penumbra widths (80%-20%) were determined. In addition, gap factors were measured with 0, 5, 20 and 40 mm distance between the collimator and the surface of the phantom. For mice irradiation, the animals were immobilized in a plastic box and positioned in direct contact with the Cerrobend block. The Farmer-type ionization chamber was placed in a polystyrene phantom under the box to enable relative output measurements that could later be correlated to dose via film measurements. In addition, EBT-XD film was used for *in vivo* measurements.

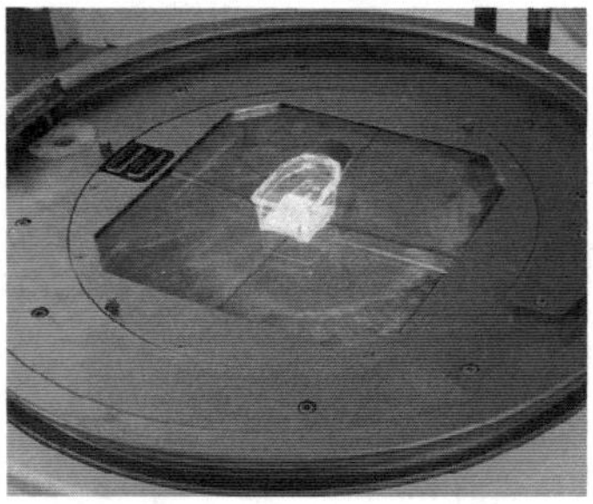

Figure 1. Cell flask with EBT^3 film and 2 mm solid water to mimic the experimental cell irradiation setup.

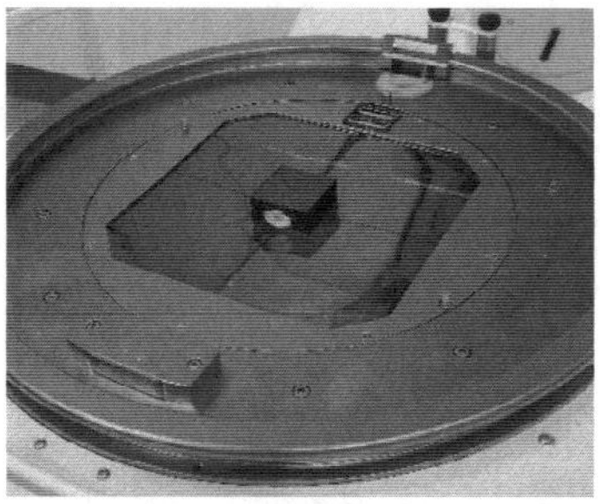

Figure 2. Experimental setup for irradiation of zebrafish embryos in an Eppendorf tube fitted into a 3D printed plastic phantom.

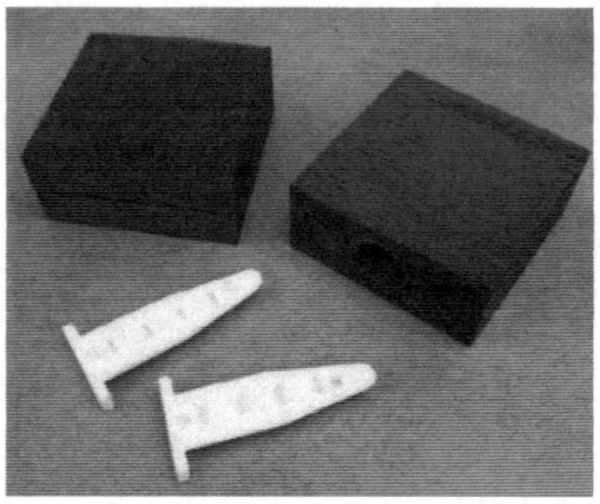

Figure 3. 3D printed phantoms together with an insert in the shape of an Eppendorf tube containing four slots for TLDs for dosimetric measurements.

3. Results and discussion

3.1. Irradiation of cells and zebrafish embryos
The dose profile on top of the cross-hair foil had a FWHM of 14.2 cm and a penumbra width of 6.8 cm. In the small area covered by the cell flask (12.5 cm^2) or 3D printed phantom (16 cm^2), the beam flatness was +/-2.2% or +/-2.7%, respectively. The absolute mean deviation between dose measured with EBT^3 film and TLD was 2.4% (range: 0.2-5.4%), which is considered acceptable given the uncertainties of these techniques. The response of the ionization chamber positioned in the ceiling was linear to the dose measured with film within the given dose range, making it useful as a relative dosimeter. The average dose delivered to the cells and zebrafish embryos were 3.1 Gy/pulse and 3.2 Gy/pulse, respectively, corresponding to average dose rates of ≥620 Gy/s and ≥640 Gy/s.

3.2. Small animal irradiations
Dose profiles and depth dose curves for each collimator are presented in Figure 4. The measured FWHMs, penumbra widths, R_{50}-values and output factors are showed in Table 1. Gap factors at distances 0, 5, 20 and 40 mm between the collimator and the phantom surface was 1.0, 0.98, 0.93 and 0.86, respectively. For a total of six *in vivo* measurements, the absolute deviation between dose measured with film and dose estimated by the Farmer-type ionization chamber reading was 2.0% (range: 0.1-4.2%). The average dose delivered to the animals was 2.0 Gy/pulse, corresponding to an average dose rate of ≥400 Gy/s.

3.3. Preparation for clinical trials
In upcoming clinical trials with veterinary patients and humans, the same electron applicator as for the small animal irradiations previously described will likely be used, with an extended selection of collimator sizes. However, before initiating clinical trials, some technical challenges must be overcome. As currently operated, interruption of the beam after the desired number of pulses is solely dependent on a diode placed in the radiation field functioning as a pulse counter. To increase the safety during ultra-high dose rate delivery, we are working on a solution where the two independent channels in a transmission monitor chamber can be used to interrupt the beam, similar to the method used for monitoring the output in conventional radiotherapy. Furthermore, it is desirable to use the transmission monitor chamber for on-line dosimetry, which is complicated due to the drop in ion collection efficiency at these high dose rates. We have previously investigated the ion collection efficiency in the built-in monitor chamber at FLASH dose rates and established an empirical model to correct for the ion recombination [14]. In future work, this model will be integrated with the accelerator to enable accurate on-line dose measurements in clinical trials.

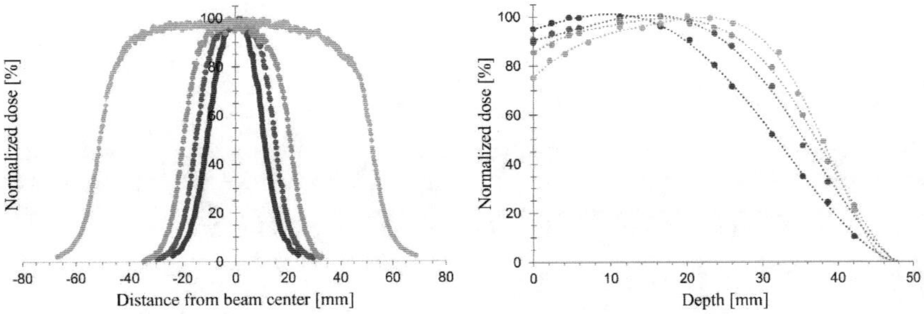

Figure 4. Dose profiles at 20 mm depth (a) and depth dose curves in solid water (b) for collimator sizes used for small animal irradiations, i.e. 10x10 cm^2 (light blue), d=4 cm^2 (light grey), d=3 cm^2 (dark blue) and d=2 cm^2 (dark grey).

Table 1. Output factor, FWHM, R_{50}-value and penumbra width at 20 mm depth in solid water, for the collimator sizes used for small animal irradiation.

	10x10 cm^2	d = 4 cm^2	d = 3 cm^2	d = 2 cm^2
Output factor	1.0	0.96	0.89	0.76
FWHM [cm]	10.3	4.1	3.0	2.0
R_{50} [cm]	3.8	3.7	3.5	3.2
Penumbra (80-20%) [cm]	1.1	0.9	0.8	0.1

4. Conclusion

In this work we have established dosimetric procedures that enable accurate dose delivery to cell cultures, zebrafish embryos and small animals at ultra-high dose rates, delivered with a modified clinical linear accelerator. We also propose a possible solution that would allow for the built-in monitor chamber to function as a dose monitor. Such a system for controlling beam-hold will allow safe delivery of FLASH radiotherapy in future clinical trials. Radiobiological experiments are currently being performed using the setups and dosimetric procedures described here.

5. Acknowledgments

This research was supported by Mrs Berta Kamprad Foundation and the Swedish Cancer Society.

6. References

[1] Favaudon V *et al* 2014 *Sci Transl Med.* **6** 245ra93
[2] Montay-Gruel P *et al* 2017 *Radiother Oncol.* **124** 365-9
[3] Vozenin M C *et al* 2019 *Clin Cancer Res.* **25** 35-42
[4] Montay-Gruel P *et al* 2019 *Proc Natl Acad Sci USA* **116** 10943-51
[5] Vozenin M C *et al* 2019 *Clin Oncol.* **31** 407-15
[6] Fouillade C *et al* 2019 *Clin Cancer Res.* CCR-19-1440
[7] Wilson J D *et al* 2019 *Front. Oncol.* **9** 1563
[8] Lempart M *et al* 2019 *Radiother Oncol.* **139** 40-5
[9] Adrian G *et al* 2019 *Br J Radiol.* **92** 1106
[10] Petersson K *et al* 2017 *Med Phys.* **44** 1157-67
[11] Boag J W *et al* 1996 *Phys Med Biol.* **41** 885-97
[12] Karsch L *et al* 2012 *Med Phys.* **39** 2447-55
[13] Jaccard M *et al* 2017 *Med Phys.* **44** 725-35
[14] Konradsson E *et al* 2020 *Radiat Res.* **194** 580-86

Journal of Physics: Conference Series **2167** (2022) 012004 doi:10.1088/1742-6596/2167/1/012004

Validation of spine SBRT using a 3D printed Anthropomorphic phantom

KH Dekker[1], KM Alexander[1], F Ynoe de Moraes[1,2], T Olding[1,2,3]
[1]Cancer Centre of Southeastern Ontario at Kingston General Hospital, 25 King Street West, Kingston, ON, K7L 5P9, Canada
[2]Department of Oncology, Queen's University, Kingston, ON, K7L 3N6, Canada
[3]Department of Physics, Queen's University, Kingston, ON, K7L 3N6, Canada

E-mail: kurtis.dekker@kingstonhsc.ca

Abstract. A segment of a spine was 3D-printed based on real patient anatomy, using metal-doped high density plastic to radiographically mimic bone. This spine was submerged in a water tank to create an anthropomorphic phantom. The spine print incorporated a slot for Gafchromic EBT3 film dosimeters and fiducials for alignment of measured and calculated dose distributions. Spine SBRT treatment plans were generated for both 6 MV and 10FFF energies based on oncologist-drawn contours transferred from real anatomy. Plans were delivered under image guidance using our clinical procedures, to evaluate the dosimetric accuracy of our planning system in high density inhomogeneities and the geometric accuracy of delivery. Results show that the Acuros XB algorithm (dose-to-water) agrees well with film measurements throughout the measured region, including within the bone substitute material. Alignment of the steep dose gradients in planned and measured doses was within 0.5 mm in the ANT-POST direction and within 0.9 mm in the SUP-INF direction, both within machine tolerances. Our results give us confidence in our ability to plan and accurately deliver spinal SBRT treatments.

1. Introduction

Stereotactic body radiation therapy (SBRT) is used to treat some patients suffering from spinal metastases [1, 2]. Spine SBRT plans exhibit very steep dose gradients in order to spare the spinal cord while delivering sufficient dose to the target volume, which is comprised of the vertebral body and often partially wraps around the cord. Our clinic has recently begun development of a spinal SBRT program. To assist in protocol development and validation, we created an anthropomorphic spine phantom that enables the use of planar dosimeters to measure dose in the target and critical structures. In this report we present the phantom design and results of our initial spine SBRT tests. The purpose of this work is to 1) validate the image guidance technique for the treatment site, and 2) investigate the accuracy of dose calculation within high-density inhomogeneities using our clinical treatment planning system. Because our aim was to measure dose within a high density object, gel dosimeters were not readily applicable to this experiment.

2. Methods

2.1. Anthropomorphic spine phantom
A segment of the thoracic spine (7 vertebrae, approximately 17 cm long) of a previously-treated lung cancer patient was contoured and exported to a 3D triangulated mesh. The mesh was processed for 3D printing using Slic3r software (Slic3r Prusa Edition) to export G-code 3D printer instructions for a fused deposition modelling 3D printer (Prusa i3 MK3, Prusa Research). Printing was performed (100% infill) with an iron-doped polylactic acid (PLA) filament (3D Printing Canada), resulting in a mean CT number of approximately 750 HU and a relative electron density of approximately 1.4 [3]. Thus, the printed spine appears similar to bone on CT / CBCT imaging, allowing us to test the accuracy of our clinical image guidance techniques for patient setup. For this study, the spine was printed in two halves, cut along the sagittal plane, creating a slot for the insertion of a single piece of film (Figure 1a). Metal BBs were embedded in the flat face of the film slot to facilitate registration of dose distributions in analysis.

The printed spine was then placed within a body-shaped water phantom (Modus QA), held in place by a custom made plastic mount made from polycaprolactone (PCL). The mount was made by heating the PCL (Polydoh, Materialix Inc.) to 60 °C, at which point it becomes easily mouldable. The spine phantom was pressed into the PCL, creating a unique imprint for reproducible positioning. This mount was attached to a 0.5 cm thick sheet of Superflab using double-sided tape. This assembly was similarly affixed to the bottom of the water tank. Figure 1b shows the complete, water-filled phantom.

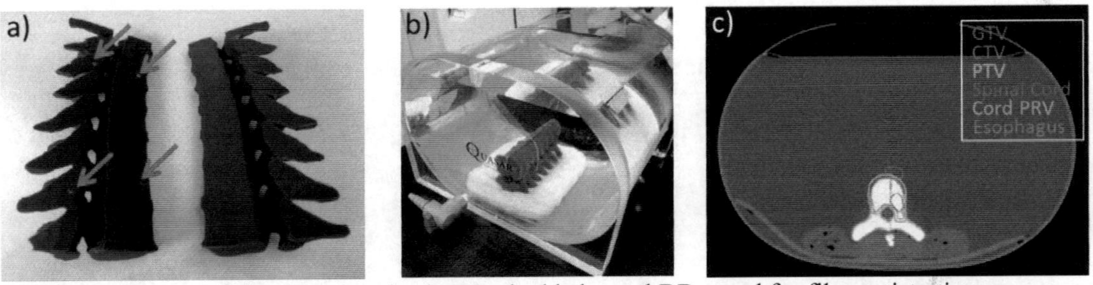

Figure 1. a) Halves of the 3D printed spine. Embedded metal BBs used for film registration are indicated by the red arrows. **b)** Assembled phantom filled with water. **c)** Axial slice of planning CT scan of the phantom, with contoured target/OAR volumes transferred from corresponding patient CT.

2.2. Treatment planning and delivery

2.2.1. Volume contouring
Realistic GTV, CTV, PTV volumes, as well as spinal cord and esophagus OARs, were contoured by a radiation oncologist on the original patient anatomy. A spinal cord PRV was generated using a 2 mm expansion on the spinal cord. For optimization and prescription purposes, a modified PTV was created by cropping the PTV out of the esophagus and cord PRV volumes. Subsequently, the contours were transferred to a planning CT scan of the phantom (Figure 1c) by rigidly registering the patient CT image to the phantom image (based on the spine).

2.2.2. Plan optimization and dose calculation
The treatment was initially planned to deliver a prescribed dose of 27 Gy in 3 fractions, normalized such that $V_{100\%} = 90\%$ for the modified PTV (volume = 20.5 cc). Additionally, the plan was optimized such that the maximum point dose within the PTV was less than 130% of the prescription, and that $V_{100\%} \geq 99\%$ for the GTV. For OARs, the dose tolerances were $D_{0.1cc} < 24$ Gy for the cord PRV, $D_{0.1cc} < 22$ Gy for the spinal cord proper, and $D_{0.1cc} < 27$ Gy for the esophagus (additionally, $D_{0.35cc} < 20$ Gy).

The treatment was planned using both a flattened 6 MV beam as well as a flattening filter free 10 MV beam (10FFF). Two coplanar full arcs (collimator 30° and 330°) were used.

The dose distribution was calculated using the convolution-based Analytical Anisotropic Algorithm (AAA) and the Acuros XB algorithm, a grid-based Boltzmann equation solver (v13.6, Varian), both with a 2 mm resolution calculation grid. Since the film dosimeter is near water equivalent, dose-to-water was determined to be the appropriate dose reporting method when comparing Acuros calculations to film measurements.

2.2.3. Treatment delivery
For each delivery, the spine phantom was assembled with a piece of radiochromic film (Gafchromic EBT3, International Specialty Products, Inc) inserted and held in place with tape. Films were marked with permanent marker at the location of each metal BB, for registration of dose distributions. The prescription dose was scaled down such that the maximum dose was approximately 3 Gy, to account for the dynamic range of the readout system. Treatment was delivered using a Varian Truebeam linear accelerator equipped with a 6 degree of freedom couch for pitch and roll corrections. The phantom was set up using laser alignment marks followed by cone beam CT with a bony match on the spine for shift and rotational corrections. Image matching was performed by radiation therapists, simulating the clinical process.

2.3. Dosimetry

2.3.1. Film readout and calibration
In this study, Gafchromic EBT3 film dosimeters were used. Films were imaged using an in-house point scanner, which operates in a raster-scan fashion to acquire images at 0.6 mm spatial resolution with very low stray light contamination [3]. Measured optical densities were converted to dose using a calibration curve obtained from a series of well characterized 10 x 10 cm^2 field irradiations (depth = 2 cm) to films from the same batch (8 dose levels between 0-4 Gy). Phantom film measurements were corrected for variation in machine output and CBCT imaging dose (approximately 2 cGy).

2.3.2. Dosimetric analysis
Analysis was performed using the Film Dosimetry Slicelet within the 3DSlicer software [4, 5] (www.slicer.org). The plane of the calculated dose distributions corresponding to the film dosimeter was extracted, aided by the fiducial markers in the phantom. This plane was aligned with the measured film using the optical fiducials corresponding to the BBs. Dose profiles were taken along the steep gradient at the mid-height of the target volume in the ANT-POST (AP) direction, as well as through the target in the SUP-INF (SI) direction, for qualitative comparison. Additionally, 2D gamma dose comparisons were performed within the film sheet, using a 3%/1mm criteria (5% dose cutoff). The tight distance criterion reflects the importance of geometric accuracy when treating targets adjacent to the spinal canal.

3. Results and Discussion
Figure 2a shows a zoomed axial view of the 6 MV calculated dose distribution (AAA). The sagittal slice corresponding to the film plane is shown in Figure 2b. Figure 2c-f shows dose profiles along the lines indicated in Figure 2b. Table 1 contains the gamma passing rates (3%/1mm, 5% cutoff, evaluated within the film sheet) for the comparison of calculated and measured doses for the 6 MV and 10FFF plans. Our results are discussed from the standpoint of both dosimetric and geometric accuracy in the next sections.

Table 1. Gamma pass rates for measured vs. planning system doses.

Plan	2D Gamma Pass Rate (3%/1mm)
6 MV AAA	85.4%
6 MV Acuros	92.1%
10FFF AAA	78.6%
10FFF Acuros	90.1%

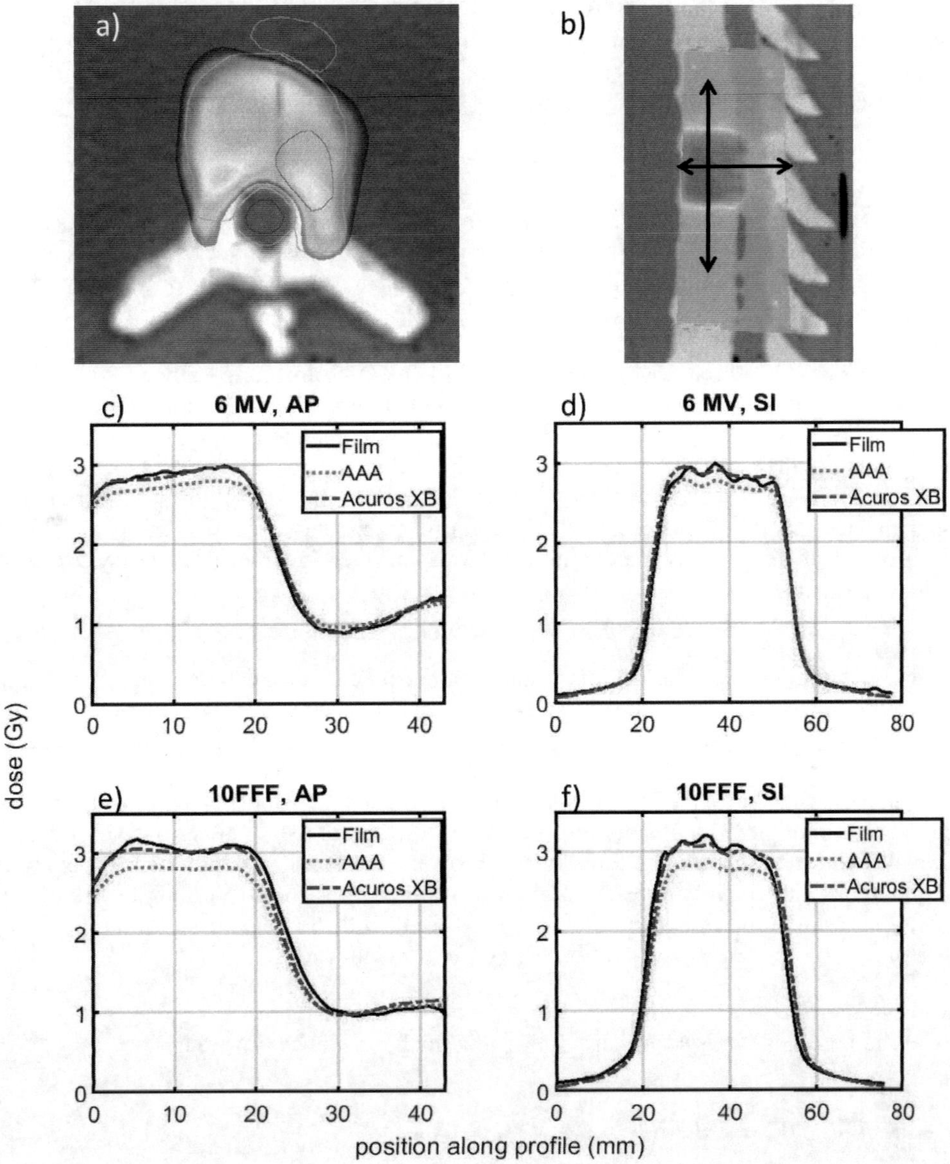

Figure 2. a) Zoomed-in axial view of the 6 MV treatment plan (AAA dose calculation), demonstrating cord sparing. Colourwash ranges from 90% of prescription to the maximum dose in the plan (122.8%). Note that the film slot can be seen on the CT image. **b)** Zoomed-in sagittal view of planning CT, with registered film dose image overlaid. **c-f)** Calculated and measured dose profiles along the lines specified in (b) for the 6 MV plan (c,d) and 10FFF plan (e,f). Note the scaling difference in the horizontal axes of the AP and SI profile plots.

3.1. Dosimetric accuracy
Examining the dose profiles in Figure 2, we clearly see that the AAA-calculated dose does not agree well with the film measurement in the high-dose region (inside the bone substitute), underestimating dose by approximately 5-7%. This is reflected in the 2D gamma comparison results (figure not shown), wherein the majority of failures lie within the target and not near the dose gradients. Meanwhile, the Acuros calculation shows very strong agreement within the high dose region. In our experiment, the film measurements correspond to a dose-to-water in "bone", while AAA reports a dose to scaled water. With Acuros, dose-to-water in medium (i.e. the dose to a small water voxel surrounded by the medium) was calculated. Therefore, it is not surprising to see much better agreement with Acuros than with AAA when using film dosimeters sandwiched in a high density object. Our discrepancy between dose-to-water (Acuros / measurement) and dose to scaled water (AAA) is consistent with the results of previous Monte Carlo simulation work by other groups [6].

3.2. Geometric accuracy
With the discussion of dosimetric accuracy in mind, we can compare the film measurements and Acuros dose calculations to evaluate the geometric accuracy of our delivery. Offline examination of CBCT-CT registrations performed by therapists showed very good alignment of the spine (Figure not shown). From the profiles (Figure 2), we see that the steep dose gradients in the SUP-INF and ANT-POST directions are very well aligned, indicating a high degree of accuracy in setup and registration along those directions. For both 6 MV and 10FFF deliveries, the location of the 50% isodose line of the steep dose gradients is within 0.5 mm in the ANT-POST direction and within 0.9 mm in the SUP-INF direction. These distances lie within tolerance for machine treatment/imaging isocentre.

2D gamma comparisons for Acuros vs. film showed pass rates above 90% for both 6 MV and 10FFF plans at 3%/1mm criteria. Considering the strict distance threshold, for a 2D comparison (less forgiving than 3D) this result is reasonable, especially given the agreement seen in dose profiles. Overall, our results indicate that we are able to achieve sufficient geometric accuracy for linac-based spine SBRT treatments.

4. Conclusion
In this study, we demonstrated the use of an anthropomorphic spine phantom for the validation of a spinal stereotactic body radiation therapy treatment protocol. We achieved very good agreement between film measurement and dose calculations performed using the Acuros XB algorithm (dose-to-water) in the high-density bone substitute. Geometrically, we observed very good alignment of the sharp dose gradients in the SUP-INF and ANT-POST directions, which gives us confidence in our image guidance technique. Our results are a strong indicator of our ability to plan and accurately deliver spinal SBRT treatments. The key limitation of our work was the lack of real patient motion, and correspondingly the inability to test immobilization techniques. Also, our initial work only utilized 2D dosimetry in a single plane. Future work will extend to 3D through the use of multiple film planes, which will allow a more complete validation of geometric accuracy.

5. Acknowledgments
We thank our radiation therapists for their input in the image guidance procedure.

6. References
[1] Ahmed K *et al.* 2012, *Int. J. Radiat. Oncol.* **82** 5, e803-9
[2] De Bari B *et al.* 2016 *Crit. Rev. Oncol. Hematol.* **98** 147-58
[3] Alexander K 2020 PhD thesis, Queen's University, Kingston, Ontario, Canada
[4] Pinter C *et al* 2012 *Med. Phys.* **39** 6332-8
[5] Alexander K *et al.* 2017, *J. Phys. Conf. Ser.* **847** 012061
[6] Bush *et al* 2011 *Med. Phys.* **38** 2208-21

Journal of Physics: Conference Series **2167** (2022) 012005 doi:10.1088/1742-6596/2167/1/012005

Monte Carlo Simulations of a Non-Invasive Positron Detector to Measure the Arterial Input Function for Dynamic PET

Liam Carroll[1], Shirin A. Enger[1,2,3,4]
[1]McGill University Medical Physics Units, Montreal, Quebec, Canada
[2]Department of Oncology, McGill University, Montreal, QC, Canada
[3]Research Institute of the McGill University Health Centre, Montreal, QC, Canada
[4]Lady Davis Research Institute for Medical Research, Jewish General Hspital, Montreal, Quebec, Canada

Email: liam.carroll@mail.mcgill.ca

Abstract. This work presents Monte Carlo (MC) study of a novel non-invasive positron detector, hereinafter called NID, designed to measure the arterial input function (AIF) through the wrist of a patient for use with dynamic positron emission tomography (PET). The goal of the study was to optimize a previously developed NID prototype, to determine its efficiency and ability to distinguish between the arterial and venous portions of the signal escaping a wrist phantom. A user code based on the Geant4 MC toolkit was developed to model the NID and a wrist phantom. The scintillator based detectors were modelled as 64 polystyrene cylinders, 0.97 mm in diameter, 10 cm long and capped with cylindrical photomultiplier tubes. The scintillator fibers were arranged in a single band around a 64.13 mm polyethylene cylinder representing a model of the wrist. Two cylinders, 2.30 mm in diameter were placed 6 mm apart, 2 mm below the surface of the wrist phantom representing the radial artery and vein. Two simulations were performed by placing 100 million decay events of oxygen-15 (^{15}O) or fluorine-18 (^{18}F), randomly distributed between the artery and vein. Visible wavelength photon tracking was enabled, and Photomultiplier tubes were simulated to collect the visible photons. Deposited energy per event and location of energy deposition were calculated. A python algorithm was used to analyse the results. The arterial signal produced a 5.28 mm and 10.32 mm FWHM for ^{15}O and ^{18}F respectively. The algorithm could determine the correct location of interaction 98% of the time. The NID can resolve the arterial and venous signal and is thus suitable for determining the AIF for dynamic PET.

1. Introduction

Clinical interest in dynamic positron emission tomography (PET) has increased in recent years[1-4]. One example of this is detection of recurrence in prostate cancer (PS). The current standard for this is to perform a gadolinium-68 (^{68}GD) labelled prostate specific membrane antigen (PSMA) PET scan. A static acquisition is performed 60 minutes post injection [5]. This protocol detects recurring lesions post treatment but is limited by the hyper-intense bladder which can mask lesions near the bladder. It has been demonstrated that dynamic PET performed up to 11 minutes post-injection, can in addition to all lesions detected with static PET also detect lesions near the bladder [6-7]. Dynamic PET has many other applications such as monitoring heart disease[8] and planning cancer treatments [3].

Unfortunately, this nuclear medicine technique is rarely used in a clinical setting due to three main disadvantages. Traditionally, dynamic PET scans are much longer than static scans, they can only acquire a small field of view and they require knowledge of the radiation concentration in the patient's arteries as a function of time, called the arterial input function (AIF). While the first two concerns are being worked on [1, 9], the acquisition of the AIF remains an unsolved problem.

The gold standard for AIF measurement is through arterial blood sampling. This is a costly and time-consuming technique that most nuclear medicine centers wish to avoid [10]. There also exists population averaging techniques [11] and image-derived input functions [12] but none have seen widespread clinical use. We are developing a non-invasive positron detector, hereinafter called non-invasive detector (NID), that can measure the AIF through the patient's wrist. This is possible due to the relatively shallow placement of the radial artery, 2 mm+/- 0.99 mm[13], compared to the maximum travel distance of common PET radioisotopes in tissue [14]. In an earlier study, a prototype of this detector was built and benchmarked against an invasive microfluidic type detector [15]. It was found that the detector was capable of measuring clinically relevant activity concentrations of ^{18}F, ^{11}C and ^{68}Ga through a wrist phantom. Figure 1 shows a regression analysis comparing the results of the two detectors. The present work presents Monte Carlo simulations to evaluate a new design for the detector to optimize its setup.

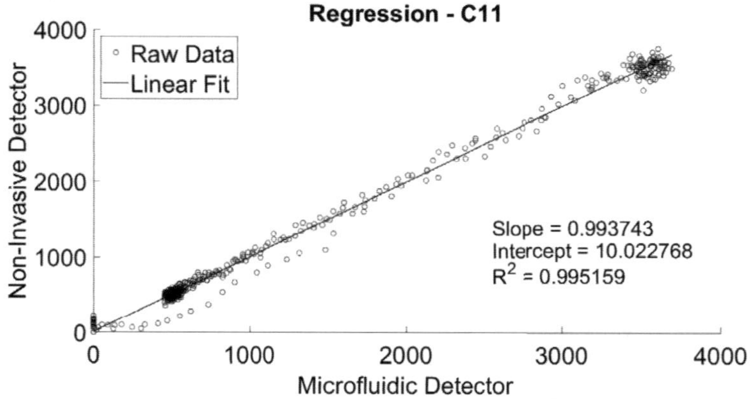

Figure 1. Regression comparing the NID to an invasive microfluidic blood sampling detector.

2. Methods

A user code based on the Geant4 MC toolkit was developed to model the NID and a wrist phantom. The NID geometry was composed of 64 plastic scintillating fibers, 0.97 mm in diameter and 10 cm long. These fibers had a 0.03 mm cladding layer and a 0.0250 mm plastic protective layer. Each fiber end was coupled to a photomultiplier tube (PMT) to collect photons in the visible range. These fibers were arranged in a single band wrapped around a 64.13 mm in diameter polyethylene cylinder that represents a wrist phantom. Two 2.30 mm diameter cylinders were placed inside the wrist to simulate the radial artery and vein of a patient. These cylinders were 6 mm apart and 2 mm below the surface of the wrist. Simulations were performed by simulating 100 million decay events of either oxygen-15 (^{15}O) or fluorine-18 (^{18}F), randomly distributed between the artery and vein. Particle transport was performed by using the Penelope low-energy electromagnetic physics and optical photon processes. Interactions were recorded when energy was deposited in the scintillator core. Total deposited energy in the scintillator was calculated and the location of interaction for each event was scored. The number of optical photons that accumulated in each PMT were also calculated. Figure 2 shows the geometry of the simulation.

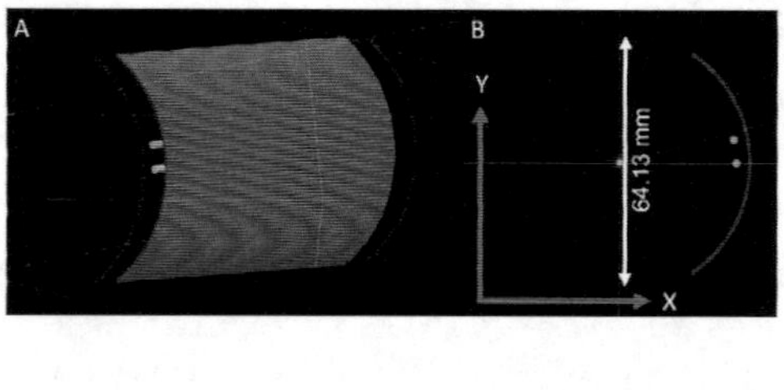

Figure 2. (A) Side and (B) front views of the Geant4 simulation.

The data was analyzed using a python algorithm to separate the arterial signal from the venous signal. For each event, the PMT with the highest photon count is considered as true location of interaction on the y-axis. The data points from positrons originating from the artery are considered and a histogram is built. A mask of this histogram is created using kernel density estimate. This mask can then be overlaid on a y-axis position histogram of a mixed arterial/venous signal which yields the arterial component of the signal. Figure 3 shows a graphical flowchart of this script.

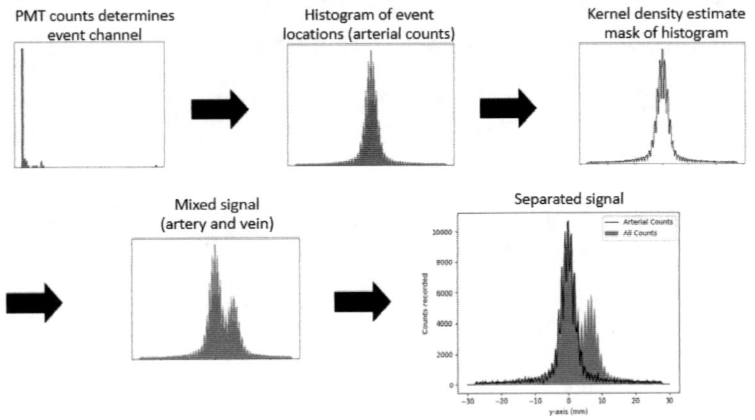

Figure 3. Flowchart of data analysis script

3. Results/Discussion

The arterial signal produced a 5.28 mm and 10.32 mm full width half max (FWHM) for ^{15}O and ^{18}F respectively. Since these two signals are 6 mm apart, for ^{15}O, this results in two distributions that are readily resolvable. For ^{18}F, the FWHM of both components of the signal are larger than their spatial separation which makes them difficult to resolve. The data analysis script takes advantage of the fact that on the first pass through the detector, the bolus will be exclusively in the patient's artery. This allows us to train the detector and use the mask at later timepoints when there will be a mixed signal. Using the script, we could calculate 104 % and 110 % of the true signal as calculated by the Monte Carlo simulation for ^{15}O and ^{18}F respectively. The number of detected positrons was overestimated by the script because the tail of the vein distribution overlaps with the arterial distribution. Currently, the script scales the mask to fit the given distribution. To improve this matching, we propose that an ultrasound of the wrist be taken before the dynamic PET scan which would allow us to measure the depth and size of the radial artery and vein. This would allow for Monte Carlo simulations to be performed on a patient-to-patient basis to match their specific biology.

4. Conclusion

We have simulated the sensitive volume of a novel non-invasive positron detector designed to detect positrons through a wrist phantom. The simulated detector can resolve the venous and arterial components of the signal and represents a promising method to non-invasively measure the arterial input function for dynamic PET.

5. References

[1] Rahmim A *et al* (2019) *Eur. J. Nucl. Med. Mol. Imaging* **46** 501-18
[2] Sachpekidis C *et al* (2019) *Cancers (Basel)* **11** 790
[3] Thorwarth D *et al* (2019) *J. Nucl. Med.* **60** 1698-704
[4] Welz S *et al* (2017) *Radiother. Oncol.* **124** 526-32
[5] Eapen RS *et al* (2019) *World J. Urol.* **37** 1255-61
[6] Sachpekidis C *et al* (2018) *Am. J. Nucl. Med. Mol. Imaging* **8** 351-9

Journal of Physics: Conference Series **2167** (2022) 012005 doi:10.1088/1742-6596/2167/1/012005

[7] Sachpekidis C *et al* (2018) *Eur. J. Nucl. Med. Mol. Imaging* **45** 904-12
[8] Petibon Y *et al* (2017) *Phys. Med. Biol.* **62** 326-43
[9] Dimitrakopoulou-Strauss A *et al* (2012) *Cancer Imaging* 12 283-9
[10] Boellaard R *et al* (2001) *Eur. J. Nucl. Med.* **28** 81-9
[11] Eberl S *et al* (1997) *Eur. J. Nucl. Med.* **24** 299-304
[12] Croteau E *et al* (2010) *Eur. J. Nucl. Med. Mol. Imaging* **37** 1539-50
[13] Lee D *et al* (2016) *J. Clin. Monit. Comput.* **30** 215-9
[14] Champion C *et al* (2007) *Phys. Med. Biol.* **52** 6605-25
[15] Carroll L *et al* (2019) *J. Nucl. Med.* **60** 384

Journal of Physics: Conference Series **2167** (2022) 012006 doi:10.1088/1742-6596/2167/1/012006

Measurement of spatial fluence distribution of neutrons and gamma rays using MAGAT-type gel detector doped with LiCl for BNCT at Kyoto University Reactor

Kenichi Tanaka[1], Tsuyoshi Kajimoto[1], Aruma Mitsuyasu[1], Yuto Ito[1], Shin-ichiro Hayashi[2], Yoshinori Sakurai[3], Hiroki Tanaka[3], Takushi Takata[3], Gerard Bengua[4] and Satoru Endo[1]

[1]Graduate School of Engineering, Hiroshima University, Higashi-Hiroshima, Japan.
[2]Faculty of Health Sciences, Hiroshima International University, Higashi-Hiroshima, Japan.
[3]Institute for Integrated Radiation and Nuclear Science, Kyoto University, Kumatori, Japan.
[4]Auckland City Hospital, Auckland, New Zealand.

Email:tanakake@hiroshima-u.ac.jp

Abstract. The spatial distributions of neutrons and gamma rays in the epithermal mode of the Kyoto University Reactor were measured using a MAGAT-type polymer gel detector doped with LiCl at ^{6}Li concentrations of 0, 10, 100 ppm. Reasonable distributions for thermal neutrons and gamma rays were obtained for 0 and 100ppm.

1. Introduction

In boron neutron capture therapy (BNCT), the quality assurance (QA) of an irradiation field can be performed by monitoring the spatial beam component distributions in a phantom. Ideally, the beam components are measured separately, e.g., thermal neutrons (<0.5eV), epithermal neutrons (0.5eV-10keV), fast neutrons (>10keV), and gamma rays. The QA regime, however, needs to be easily implemented as a routine verification method. A good candidate detector for spatial distribution measurements are a gel detector [1] which have been used in three-dimensional dose distribution measurement for photons and heavy ion irradiations [2]. In a related study, we have demonstrated that gel detectors can separately measure the relative fluence of the components of a BNCT neutron field [3]. We found from Monte Carlo calculations that the MAGAT-type polymer gel detector doped with ^{6}Li at concentrations of 0 (for fast neutron component), 10 and 100 ppm are potentially usable for thermal and fast neutrons, and gamma ray measurements. The present paper describes an experimental demonstration of these measurements using the gel detectors.

2. Materials and methods

The beam components of a BNCT irradiation field can be measured separately by increasing the sensitivity of the gel detector to the fast neutron component via recoil protons produced in the gel detector and increasing the sensitivity of the detector to the thermal neutron component via the secondary particles from the ^{6}Li(n,α)^{3}H reaction. The gamma ray component of the beam deposits its energy mainly via secondary electrons. It will therefore be possible to detect the enhanced components with the gel detector.

We previously proposed [3] potential combinations of the gel detector compositions for converters used in the standard epithermal neutron irradiation mode at the Kyoto University Reactor Heavy Water Neutron Irradiation Facility (KUR-HWNIF) which is currently used for BNCT treatments [4]. Simulations by using the Monte Carlo code PHITS ver. 2.82 [5] showed that the MAGAT-type polymer gel detector doped with ^{6}Li at concentrations of 0, 10 and 100 ppm were viable for the measurement of thermal and fast neutrons, and gamma rays. In the study, the MAGAT-type polymer gel detector was assumed to be made up of methacrylic acid, gelatin and an oxygen scavenger (tetrakis-hydroxymetyl-phosphonium chrolide, THPC). Its elemental composition was H:10.5 wt%, C:9.5 wt%, N:1.4 wt%, O:77.7 wt%, P :0.4 wt%, Cl:0.5 wt%.

In the experiments for the present study, the MAGAT-type gel detector was infused with LiCl, where the naturally abundant isotope ^{6}Li was used. The ^{6}Li concentrations were set at 0, 10 and 100 ppm. The dimension of the gel detector was 60 × 60 × 60 mm^3. The gel detector was encased in a box made of 5 mm thick acrylic acid resin. The box was set inside a 200 × 200 × 200 mm^3 acrylic acid resin phantom, to simulate a human head in a brain tumor treatment. During the irradiation, the box was placed in contact with the collimator aperture of the KUR-HWNIF. The irradiation was performed with the standard epithermal neutron irradiation mode at 1 MW delivered for 40 minutes. The aperture of the collimator was 120 × 120 mm^2. The central axes of the phantom, gel detector, and collimator aperture were aligned. The nominal value of the flux at the center of the collimator aperture was 7.07×10^{6} cm^{-2}s^{-1}, 1.33×10^{8} cm^{-2}s^{-1} and 1.38×10^{7} cm^{-2}s^{-1} for thermal, epithermal and fast neutrons, respectively. The gamma ray flux was 1.25×10^{7} cm^{-2}s^{-1}. After the irradiation, the transverse relaxation rate was measured as the signal intensity (S_i) in the unit 's^{-1}' of the gel detector, by using a 0.3T MRI scanner (AIRIS II comfort, Hitachi Medical Corp.) with a standard head coil. Here, i indicates the ^{6}Li concentration in the gel detector wherein 1, 2 and 3 denote 0, 10, and 100 ppm, respectively.

The fluence Φ_j in the unit 'cm^{-2}' of each beam component was determined using the following model:

$$D = \begin{pmatrix} S_1k_1 \\ S_2k_3 \\ S_3k_3 \end{pmatrix} = \begin{pmatrix} a_{11} & a_{12} & a_{13} \\ a_{21} & a_{22} & a_{22} \\ a_{31} & a_{32} & a_{33} \end{pmatrix} \begin{pmatrix} \Phi_1 \\ \Phi_2 \\ \Phi_3 \end{pmatrix} = A \cdot \Phi \tag{1}$$

$$\Phi = A^{-1} \cdot D \tag{2}$$

where k_i denotes the conversion factor from the signal intensity to the absorbed dose in the unit 'Gy s', where a_{ij} denotes the sensitivity of the ith gel detector for the component j and where j corresponds to the beam components, i.e., 1 for gamma-rays, 2 for thermal neutrons and 3 for fast neutrons. The sensitivity is the signal intensity per unit of the beam fluence in the unit 'Gy cm^2'. The value D=S_ik_i corresponds to the absorbed dose.

The sensitivity a_{ij} was estimated using the PHITS simulation of the irradiation condition at KUR-HWNIF. Assuming that the signal intensity is proportional to the energy deposition, the ratio of the energy deposition to the beam component fluence at the gel detector regions of 5 × 5 × 5 mm^3 along the central axis was computed and assigned as a representative of a_{ij}. The conversion factor k_i from the signal intensity to the absorbed dose was measured for the irradiation with ^{60}Co gamma rays at the Graduate School of Advanced Science and Engineering, Hiroshima University [6].

As a reference, the fluence of thermal neutrons and gamma rays was measured with gold wire activation combined with ^{nat}Cd covers, and thermoluminescence dosimeter (TLD), respectively. For the TLD measurements, the dose was assumed to be proportional to the fluence. The gold wire and TLDs were placed on the central axis of the gel detector and their irradiations were performed separately from the irradiation to the gel detectors.

3. Results and discussion

The sensitivity of the gel detector, a_{ij}, was calculated with the PHITS code and is shown in figure 1. The depth shown is the distance from the surface of the gel detector set inside the box made of 5 mm thick acrylic acid resin. The relative contributions of the beam components to the total of the energy depositions in the gel detectors are shown in figure 2. The uncertainties in figures 1 and 2 were estimated by considering only statistical uncertainty by the PHITS simulation. The sensitivities in figure 1 were used for the fluence estimation with the equation (2).

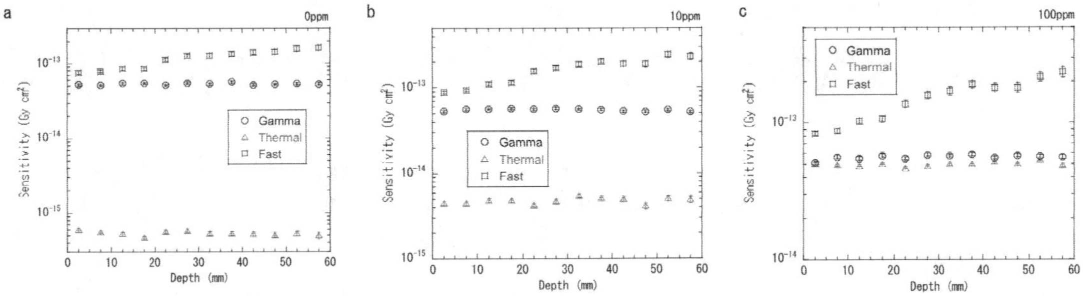

Figure 1. Sensitivity (the ratio of energy deposition to the beam component fluence) of the gel detector with ^{6}Li at (a) 0 ppm, (b) 10 ppm, and (c) 100 ppm.

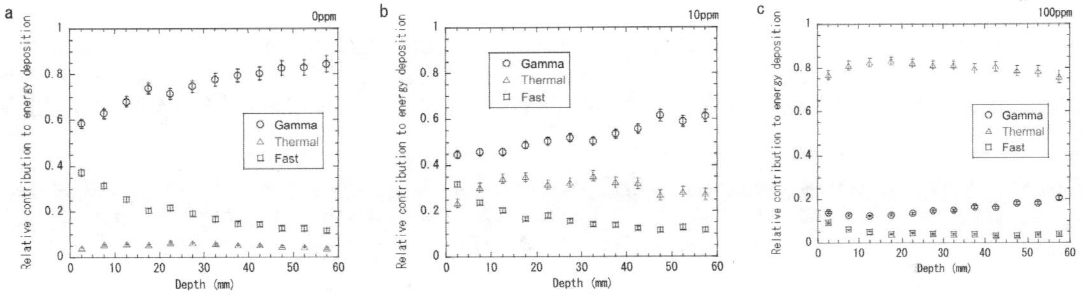

Figure 2. Contribution of beam component to energy deposition in the gel detector with ^{6}Li at (a) 0 ppm, (b) 10 ppm, and (c) 100 ppm.

The raw signal intensity of the gel detector is shown in figure 3 (a). The plotted value and error bar shown are the average and standard deviation of the signal intensity in the region of $5 \times 5 \times 5$ mm^3 along the central axis of the gel detector. The background signal intensity, without the irradiation, was 6.3, 6.1, and 5.1 s^{-1} for the gel detector with LiCl at ^{6}Li concentrations of 0, 10, and 100 ppm, respectively. As a result of the irradiation with the ^{60}Co gamma rays, the signal intensity dependence on the absorbed dose is 7.2, 7.5, and 7.1 s^{-1}Gy^{-1}, at ^{6}Li concentrations of 0, 10, and 100 ppm, respectively. These values were treated as the dose-response of the gel detectors.

The data in figure 3 (a) was corrected so that all the gel detectors would have the same dose-response as the gel detector without ^{6}Li, and the background signal intensity was subtracted. The corrected result is shown in figure 3 (b). The data at the depths of 2.5 and 57.5 mm, which are within the regions 0 to 5 mm away from the wall of the acrylic acid resin box, were omitted from further analysis since the oxygen penetrating the wall reduced the polymerization and thus the signal intensity.

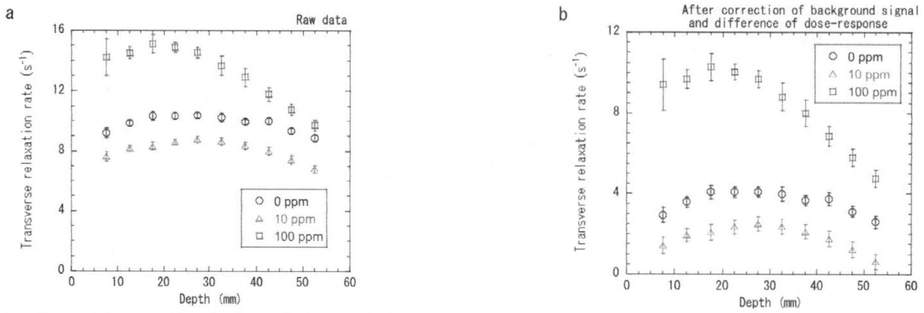

Figure 3. Signal intensity (s^{-1}), of the gel detector; (a) raw data, (b) after correction of background signal and difference of dose-response.

The fluence estimated using the results for the three gel detectors is shown in figure 4. The distributions are normalized to the maximum value. Part of three beam components yielded negative values. The calculated relative contribution of fast neutrons to the total of the energy deposition in the gel detectors, which is shown in figure 2, is about 30% near the gel detector surface for 0 and 10 ppm, and is below 20% for other depths. It is only a few percent for all the depth for 100 ppm from figure 2 (c). As a trial, the contribution of the fast neutron component was not included or corrected, and only gamma rays and thermal neutrons were computed without using the gel detectors having 10 ppm ^{6}Li.

The results for the two-component estimation are shown in figure 5. In this case, both the gamma rays and thermal neutrons are positive. The standard deviation is about 2% for gold activation, 10% for TLD, and 0.5% for the PHITS calculation. For clarity of plotted data, the error bars for those data are not shown. The fluence distribution measured by the gel detector agrees to within 20% to 30% of other estimates. This suggests the potential usability of polymer gel detector in spatial measurement of fluence in BNCT beam.

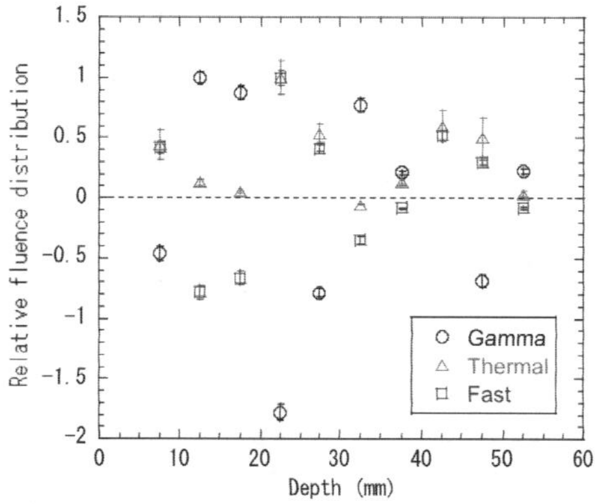

Figure 4. Relative distribution of fluence estimated with three gel detectors.

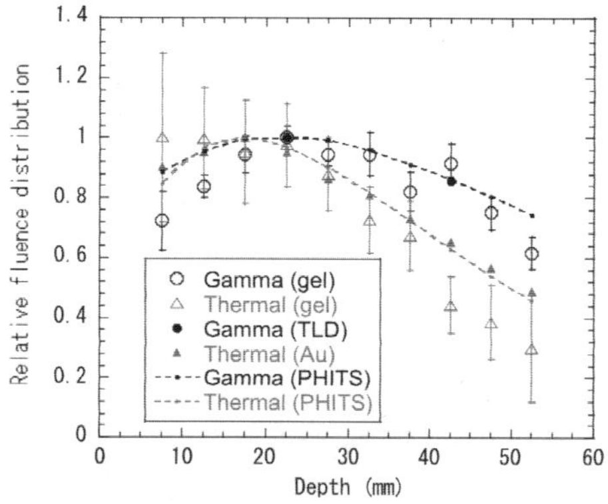

Figure 5. Relative distribution of fluence estimated with two gel detectors with ^{6}Li at 0 and 100 ppm.

4. Conclusion

A trial to measure the spatial distribution of the neutrons, depending on their energy, and gamma rays separately was carried out by using the MAGAT-type gel detector infused with LiCl. Potential usability of the gel detector was shown for two components, i.e., thermal neutrons and gamma rays, for which the plausible fluence distributions was obtained. This suggests the potential usability of polymer gel detector in spatial measurement of fluence in BNCT beam.

5. Acknowledgments
This work was supported by JSPS KAKENHI Grant Numbers JP26293281. This work has been carried out in part under the visiting Researcher's Program of the Institute for Integrated Radiation and Nuclear Science, Kyoto University. The authors express their sincere appreciation to Mr. Yuto Murakami and Mr. Yuto Ito in Hiroshima University, and the Innovation Plaza in Hiroshima University (especially to Mr. Seisuke Noguchi), Japan for their support in the investigations.

6. References
[1] Khajeali A, Farajollahi AR, Kasesaz Y *et al* 2015 *Appl. Rad. Isot.* **105** 72-81
[2] De Deene Y and Vandecasteele J 2013 *J. Phys. Conference Series* **444** 012015
[3] Tanaka K, Sakurai Y, Hayashi S *et al* 2017 *Appl. Rad. Isot.* **127** 253-259
[4] Sakurai Y and Kobayashi T 2000 *Nucl. Instr. Meth.* **A453** 569-596
[5] Sato T, Niita N, Matsuda N *et al* 2013 *J. Nucl., Sci. Technol.* **50** 913-923
[6] Murakata A, Endo S, Kojima Y *et al* 2010 *J. Radiat. Res.* **51** 198-203

Journal of Physics: Conference Series **2167** (2022) 012007 doi:10.1088/1742-6596/2167/1/012007

Development of a practical clinical application of NIPAM kV-CBCT dosimetry

K Pant[1], M Oldham[2], W Giles[2] and J Adamson[2]
[1] Medical Physics Graduate Program, Duke University, Durham, NC 27708, USA
[2] Department of Radiation Oncology, Duke University Medical Center, Durham, NC 27708, USA

Email: justus.adamson@duke.edu

Abstract. We report our progress towards developing a clinical application of NIPAM kV-CBCT dosimetry. The goal is to develop a practical kV-MV isocenter verification test for which the measurement and analysis can be carried out quickly (within an hour), and that eliminates the need for separate readout (other than on board kV-CBCT) or extra analysis steps such as image registration. Isocenter verification is performed using a NIPAM 3D gel dosimeter which is irradiated with a small field to ~16Gy at eight unique couch/gantry angles. Pre- and post-irradiation kV-CBCT images are acquired and dose is manifest as the intensity difference between pre- and post-CBCTs due to radiation induced changes in density. Code was developed to detect the geometry of each beam in the kV-CBCT and quantify relevant parameters. We applied this technique to verify the isocenter for MLCs as well as for SRS cones. The measured radius to encompass all beams for 4mm, 6mm, 7.5mm, 12.5mm, and 15mm cones was 0.55±0.11mm. The efficiency, robustness to setup errors, and unique ability to visualize spatial uncertainties in the kV-CBCT coordinate system make the NIPAM kV-CBCT test a practical and unique tool for kV-MV isocenter verification.

1. Introduction

Advances in patient imaging have significantly improved the workflow of treating patients with stereotactic body radiation therapy (SBRT) and stereotactic radiosurgery (SRS). A major breakthrough in patient imaging is the use of a kilovoltage (kV) imaging system attached to the linear accelerator that is capable of acquiring cone-beam computed tomography (CBCT) images [1]. This is especially useful during SBRT/SRS for the verification of patient alignment [2] and the evaluation of organ motion [3] throughout the course of treatment. On-board kV-CBCT is also increasingly replacing external stereotactic frame systems, which require fixing the unit to the patient's skull, by aligning a frameless mask using the imaging system [4].

Because of the critical clinical roles in which the kV-CBCT system takes part during patient treatment, the spatial accuracy of the radiation and imaging system is of high importance. This is tested and verified at regular intervals depending on the utilization of the linear accelerator [5]. Three-dimensional (3D) dosimetry systems may offer more comprehensive QA tests and provide information about the dose distributions achieved by techniques such as SBRT/SRS [6]. In this study, we demonstrate the use of kV-CBCT dosimetry to verify the spatial accuracy of the kV-MV isocenters. kV-CBCT dosimetry is enabled by N-isopropylacrylamide (NIPAM) polymer gel dosimeters. These are fabricated from radiation sensitive materials; when irradiated, a change in density occurs due to polymerization [7]. Visualization of the dose is made possible due to the change in density also causing a change in intensity [8] in the kV-CBCT image. Specific formulations of NIPAM gel dosimeters have been optimized for x-ray CT dosimetry, and are useful due to their reduced toxicity [9] and high sensitivity for x-ray CT [10]. Previous work [11] has demonstrated the use of NIPAM dosimeters for comprehensive verification of the radiation isocenter uncertainty and coincidence with the kV-CBCT imaging system using only three collimation systems.

Our purpose here is to report our progress towards developing a practical kV-MV isocenter verification test using NIPAM kV-CBCT dosimetry. We also apply this test for various clinical scenarios, including verification of all SRS cones, as well as before and after scheduled isocenter alignment maintenance of a Varian TrueBeam STx.

2. Materials and Methods

2.1. Isocenter verification test

NIPAM dosimeters were prepared at Duke University using an established formula [8]. The gel was poured into 16oz. Polyethylene Terephthalate (PET) plastic cylindrical jars (7.5cm diameter and 13.1cm height). To minimize oxygen effects, mineral oil was poured on top of the gel to displace oxygen and the top covered with parafilm prior to capping. The dosimeters are placed in a refrigerator to set (6-8 hours) and stored there until ready for use.

The isocenter verification test begins by immobilizing the dosimeter on the treatment table. In our initial tests, immobilization was accomplished by placing the dosimeter in a make-shift Styrofoam apparatus, and securing it further with masking tape. However, this was later replaced with a more secure thermoplastic immobilization in order to increase the spatial accuracy of the test. A CBCT image is then acquired. A slow-rotation CBCT scan is used to increase the number of projections from 895 (in a standard scan) to 5,330. The following settings were used: 125 kVp, 5130 mAs, 1mm slice thickness, smooth reconstruction. The dosimeter is then irradiated at eight unique gantry/couch angles, chosen specifically so that they do not overlap except at the isocenter. A post-irradiation CBCT is immediately acquired using the same settings as the pre-irradiation CBCT. All DICOM files are then exported to Matlab for analysis.

The Matlab code determines beam geometry using an iterative optimization process that maximized the contrast-to-noise ratio (CNR) of the high-dose regions relative to the low-dose (background). The CNR is calculated using a contrast and background weighting for each voxel of the CBCT image.

2.2. Clinical applications

The isocenter verification test was previously reported for three collimation systems: 4mm cone, 7.5mm cone, and 10mm diameter MLC field [11]. Here we perform the test on the remaining SRS cone sizes (6mm, 10mm, 12.5mm, and 15mm). Both automatic and manual analysis are performed using the Matlab code developed in house.

A second clinical application of the isocenter verification test is to detect kV-MV isocenter misalignment due to machine drift. Prior to maintenance of the isocenter alignment on a Varian TrueBeam STx, the comprehensive isocenter verification test was performed using 10mm diameter MLC fields. After maintenance was complete, the test was again performed using 5mm MLCs. For this specific test, the conformal thermoplastic radiotherapy mask was used (as demonstrated in Figure 3). The results were compared with the ISO lock measurements provided by the vendor.

3. Results and Discussion

3.1. Isocenter verification test

Figure 1 is a photograph of various dosimeters which illustrates the irradiation geometry for various collimation systems. Setup, irradiation, and imaging using the slow-rotation CBCT scan could be completed in approximately 50 minutes. While the slow scan does increase the number of projections, it also greatly increases the time required for image acquisition (6 minutes per scan). Figure 2 shows the axial slice of the pre-irradiation CBCT, post-irradiation CBCT, and subtracted CBCT with a mask applied for the isocenter verification test performed using the 5mm diameter MLC field.

Figure 1. Unirradiated dosimeter (left) and irradiated dosimeters with increasingly larger field sizes.

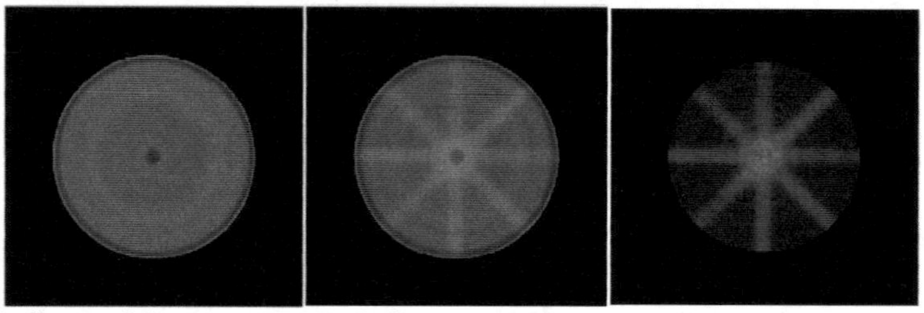

Figure 2. Axial slices of the pre-CBCT (left), post-CBCT (middle), and subtracted CBCT with a mask applied (right) using 5mm diameter MLC field.

The creation of a custom thermoplastic immobilization better secured the NIPAM dosimeter as demonstrated in Figure 3. Dosimeter shift is more pronounced on the outer edge of the dosimeter immobilized in the Styrofoam apparatus when compared to the thermoplastic immobilization. Figure 3 also shows the improved image quality of slow-rotation CBCT versus a standard CBCT, due to increasing the number of projections.

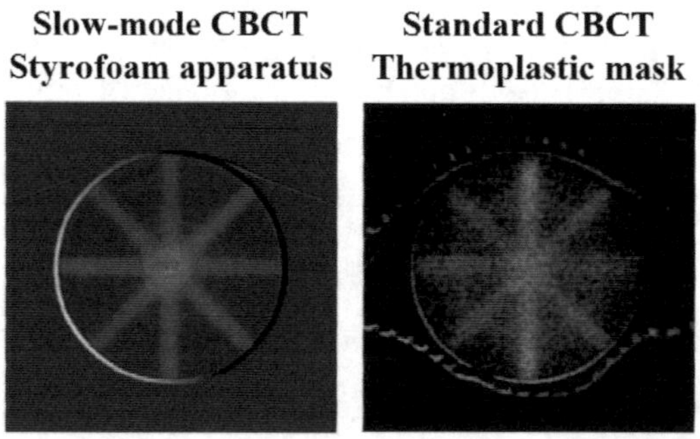

Figure 3. Subtracted (pre-CBCT from post-CBCT) images of the NIPAM dosimeter. The edge artifact indicates dosimeter shift during irradiation, which is less pronounced with thermoplastic immobilization (right). Improved image quality for slow-rotation CBCT (left) is also apparent.

3.2. Clinical applications
Table 1 shows the results of the automatic analysis for the SRS cones performed in this report. In comparison, traditional star shots indicated a minimum tangent circle radius of 0.28mm for the same machine. Winston-Lutz images with MLCs indicated a 3D isocenter radius of 0.24mm, which also matched the isocenter size determined by the Varian MPC.

Table 1. Automatic analysis results of the isocenter verification test.

Cone diameter	6mm	10mm	12.5mm	15mm
Dist. radiation iso to CBCT origin (mm)	0.23	0.15	0.65	0.14
Smallest radius intersecting all beams (mm)	0.53	0.75	0.5	0.57
Gantry accuracy mean differences (°)	0.38±0.42	0.56±0.7	0.86±0.74	0.47±0.39
Couch accuracy mean differences (°)	0.01±0.45	-0.08±0.66	-0.29±1.53	-0.07±0.55

Prior to maintenance, the machine isocenter diameter measured by the vendor was 0.976mm. After maintenance, this value decreased to 0.519mm. Our results using the NIPAM isocenter verification test with a 10mm MLC field before maintenance showed the radial distance from the radiation isocenter to the CBCT origin was 0.45mm. After repair, the isocenter verification test with 5mm MLCs showed this value to be 0.31mm. The smallest radius to intersect all beams decreased from 0.46mm to 0.43mm before and after maintenance, respectively.

4. Conclusion
This work demonstrates the clinical applications of using the comprehensive isocenter verification test with NIPAM kV-CBCT dosimetry. The fast acquisition and analysis, unique ability to directly visualize dose in the CBCT image, and comprehensive analysis of radiation uncertainty, coincidence with the kV-CBCT imaging coordinate system, and couch and gantry angle accuracy for all collimation systems, make NIPAM kV-MV isocenter verification a unique tool for a clinical radiosurgery program.

5. References
[1] Jaffray D, Siewerdsen J, Wong J *et al* 2002 *Int. J. Radiat. Oncol.* **53** 1337-49
[2] Smitsmans M, De Bois J, Sonke J *et al* 2005 *Int. J. Radiat. Oncol. Biol. Phys.* **63** 975-84
[3] Adamson J and Wu Q 2009 *Int. J. Radiat. Oncol. Biol. Phys.* **75** 260-67
[4] Dhabaan A, Schreibmann E, Siddiqi A *et al* 2012 *J. Appl. Clin. Med. Phys.* **13** 215-25
[5] Klein E *et al* 2009 *Med. Phys.* **36** 4197-212
[6] Schreiner L 2015 *J. Phys. Conf. Ser.* **573** 012003
[7] Baldock C, De Deene Y, Doran S *et al* 2010 *Phys. Med. Biol.* **55** **R**1-63
[8] Maynard E, Hilts M, Heath E *et al* 2017 *Med. Phys.* **44** 736-46
[9] Senden R, De Jean P, McAuley K *et al* 2006 *Phys. Med. Biol.* **51** 3301-14
[10] Adamson J, Carroll J, Trager M *et al* 2019 *Int. J. Radiat. Oncol.* **103** 1271-9
[11] Pant K, Umeh C, Oldham M *et al* 2020 *Med. Phys.* **47** 927-36

Journal of Physics: Conference Series 2167 (2022) 012008 doi:10.1088/1742-6596/2167/1/012008

Dose rate effects in radiochromic leuco crystal violet dosimeters

Kevin Jordan[1,2,3]
[1]Department of Medical Biophysics, Western University, London, ON, Canada
[2]Department of Oncology, Western University, London, ON, Canada
[3]London Regional Cancer Program, London Health Sciences Centre, London, ON, Canada

Email: kevin.jordan@lhsc.on.ca

Abstract. Radiochromic dosimeters based on leuco crystal violet (LCV) in aqueous solutions and gelatin hydrogels were evaluated for dose and dose rate performance. Optical transmission measurements were performed on samples in custom 10 cm long cuvettes in order to have measurable signals below 10 Gy. Standard gel formulations with 0.7 mM concentration of anionic surfactant sodium dodecyl sulfate (SDS) had a dose rate effect of increasing sensitivity with increasing dose rate of nearly 15% from 1 to 10 Gy/min at 20 Gy. Solutions of LCV and trichloracetic acid also showed similar dose rate effects. Adding LCV increased dose rate effect. Adding SDS, lowered sensitivity but maintained dose rate effect. Adding 0.4% gelatin also lowered sensitivity but reversed dose rate effect. For gels, lowering the pH reduced both sensitivity and dose rate effect. These trends are consistent with LCV solubility contributing to dose rate effects. These data indicate there are several factors related to dose rate effects in leuco crystal violet dosimeters and formulations designed to maximize sensitivity may not minimize dose rate effects.

1. Introduction

Radiochromic gelatin hydrogels with leuco dyes are one class of 3D dosimeter with several advantages for optical computed tomography (CT) readout of samples greater than 10 cm in diameter. These advantages include: low cost and non-toxic chemicals, easy preparation in open beaker environment, low scatter, low initial colour, low or very low diffusion rates. Initial results indicated gelatin hydrogels with leuco crystal violet (LCV) trichloroacetic acid (TCAA) and non-ionic surfactant Triton X100 was a promising alternate to ferrous-xylenol orange gels for larger volumes because of higher initial transmission and more stable dose image [1]. However, further experiments indicated dose rate effects which would limit conditions for accurate 3D dosimetry [2]. Further experiments in-house found a ~10% fading effect that in certain conditions could be interpreted as an effective dose rate effect. Other groups have found no dose rate effect for this initial formulation [3-5]. Recent large volume samples are measuring a dose rate effect of nearly +15% for increasing dose rate from 1 to 10 Gy/min. Why dose rate effects were not observed previously but are present in recent data is an ongoing investigation. Substituting for nonionic Triton X100 with anionic surfactant sodium dodecyl sulfate (SDS) at 0.7 mM concentration results in a low scatter, non-diffusing, 3D, radiochromic dosimeter [6]. This standard formulation: provides better agreement with known depth doses at short post gelation times than for 3D samples stored at 5 °C for several days. It also has a dose rate effect of near +15% for 12 MeV electrons at 20 Gy from 1 to 10 Gy/min and under-responds in low dose regions of arc delivery treatments. While, 12 MeV electron depth dose and repeat irradiations of small, 1 cm, cuvettes both show linear dose responses to >40 Gy, repeat irradiations of larger volume samples and 3D reconstructions reveal dose responses dependent on dose rates of previous and current deliveries. These sequential 3D irradiations are a sensitive test of combined effects of dose and dose rate in the same volumes as intended unknowns.

Because of the mixed results in the literature and current problems with this SDS formulation a set of experiments were conducted to re-examine dose rate effects. These include, examining solutions and gel formulations in custom 10 cm long by 2.2 cm diameter polymethyl methacrylate (PMMA) cuvettes. The longer cuvettes allow accurate transmission measurements for lower doses and the larger volumes may be more representative of results in the bulk of 3D dosimeters away from the vessel walls.

2. Methods

Leuco crystal violet (LCV), trichloroacetic acid (TCAA), sodium dodecyl sulfate (SDS) and gelatin (porcine, type A, 300 bloom) were purchased from Sigma-Aldrich. The order in which samples are added can produce different results. In general, the water or gel solutions are lowered in pH before adding SDS or LCV. This minimizes initial colour change. Dispersing SDS before addition of LCV results in lower scatter samples. Concentrations similar to current optimum gel formulation (5% gelatin+32 mM TCAA +1 mM LCV +0.7 mM SDS), were starting values.

The samples were usually prepared a few hours prior to irradiation and maintained at 20 °C throughout experiments. Photochromic tests indicated no measurable change in transmission for the light exposures of this study. Custom 10 cm long PMMA cuvettes were prepared by cutting fill/cleaning slots along lengths of 2.5 cm OD plastic tubing and welding 0.15 cm thick PMMA plates to the ends. After filling, samples were sealed with clear 3M, Scotch™ tape. Transmission measurements were performed with a Hitachi-Perkin Elmer model 204 absorption spectrometer at 590 nm with 1 mm slits and a water reference. Linear attenuation coefficients were calculated for direct comparisons with optical CT reconstructions. Irradiations were performed with a Varian 2100 IX series linear accelerator; 6 MV, 6 Gy/min, jaw size 20x20 cm or 12 MeV at 1 or 10 Gy/min, 4x4 or 20x20 cm cutouts. For dose rate experiments, cuvettes were irradiated as two groups in a water tank with cuvette axes at 2.8 cm depth. Irradiation and readout required 20 minutes between subsequent dose points.

3. Results

Even and odd numbered samples were irradiated at 10 and 1 Gy/min dose rates respectively. In figure 1, the stock solution was 1 mM LCV + 25 mM TCAA and S0 and S1 are same solution irradiated at 10 vs 1 Gy/min. The plot demonstrates a dose rate effect which increases with dose and has greater sensitivity for higher dose rate similar to the standard gel formulation. S10 and S11 contain stock solution + 0.7 mM SDS. This concentration of SDS lowered dose sensitivity by 50% and maintains the trend in dose rate effect. Note, 0.7 mM is below the critical micelle concentration of SDS in water. Adding 0.4% gelatin to stock solution (S40, S41) reduced sensitivity by ~65% but reversed the trend in dose rate effect. Specifically, the lower dose rate is ~5-10% more sensitive. All three samples are sublinear with dose response. S50 combines SDS with gelatin, resulting in an intermediate sensitivity but increased linearity in dose response. No sample S51 was available for comparison. These results indicate dose rate effects are dose dependent and have multiple causes.

In figure 2, the dose range was increased to 60 Gy. S4 and S5 curves show solution sensitivity at 30 Gy increased 30% by increasing LCV concentration from 1 to 2 mM, see figure 1 S0 and S1. Also the dose rate effect increased with more LCV in solution. Sublinear dose response was similar for 1 and 2 mM solutions. Increasing the TCAA concentration to 35 mM, lowered the dose sensitivity and reversed the dose rate effect below 30 Gy compared to S0 and S1. At higher doses, the dose rate effect was minimal. These results indicate solutions with LCV concentration <1mM and TCAA concentration >25 mM may have smaller dose rate effects at doses <30 Gy. In contrast, a gel formulation, S90 (4% gelatin + 1 mM LCV + 0.7 mM SDS + 25 mM TCAA), has an intermediate dose sensitivity which is linear to >60 Gy.

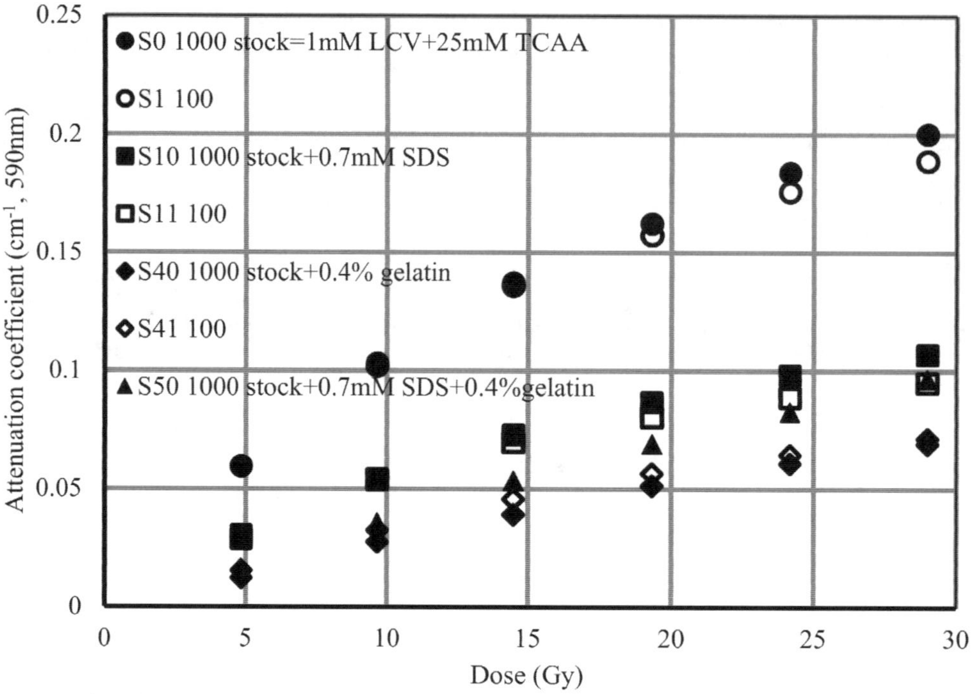

Figure 1. Plot of solution attenuation coefficient vs dose in 10 cm cuvettes with repeat irradiations. Solid/open markers at 10/1 Gy/min respectively. Reference, 0 Gy sample.

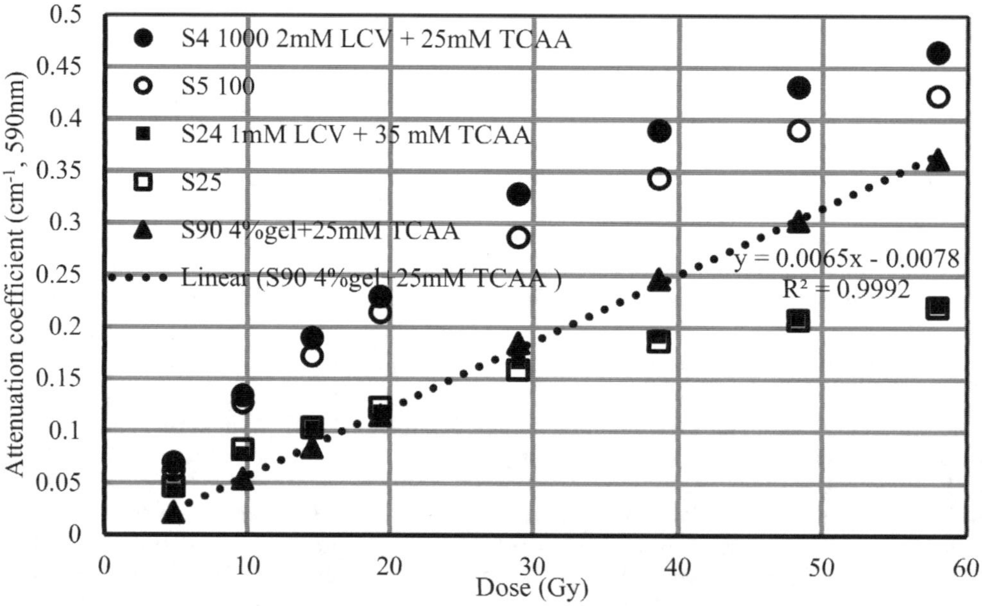

Figure 2. Plot of solution and gel attenuation coefficient vs dose in 10 cm cuvettes with repeat irradiations. Solid/open markers at 10/1 Gy/min respectively. Reference, 0 Gy sample.

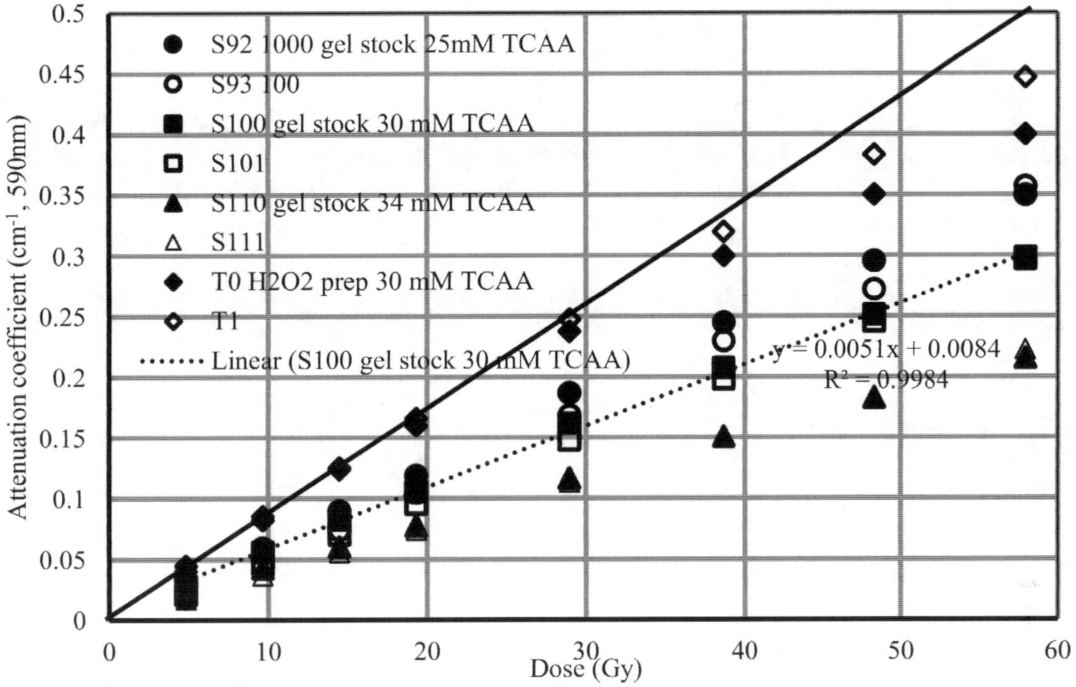

Figure 3. Plot of gel attenuation coefficient vs dose in 10 cm cuvettes with repeat irradiations. Solid/open markers at 10/1 Gy/min respectively. Reference, 0 Gy sample, dotted line, linear fit to S100 gel, solid line drawn as a visual guide.

In figure 3, the effects of increasing the TCAA concentration from 25 to 30 and 34 mM in 4% gelatin gel is shown. Note, samples S92 and S93 are from same batch as S90 but two days older at irradiation. Essentially, the sensitivity and dose rate effect decrease with increasing TCAA concentration. This trend suggests that LCV may not be completely dissolved at TCAA concentration of 25 mM in 4% gelatin or 32 mM in 5% gelatin (common formulations). Also shown, is another formulation, T0 and T1 where the gelatin solution is pretreated with 1 mM hydrogen peroxide. This step increases the initial gel dose sensitivity threefold. However, after several days of storage at 5 C the sensitivity decreases to similar levels found without this step. Note the dose rate effect for this same day sample is reversed. This data supports the potential complexity of gelatin hydrogel dosimeters.

4. Conclusion

Comparisons of solution and gel dosimeters for same chemistry provides additional information for development of hydrogel dosimeters. Ten cm long cuvettes provide more accurate measurement of dose response below 10 Gy. However, direct comparison with performance in large volume (10 cm diameter or larger vessels) is necessary before relevance of data can be determined. Other similar, in-house, experiments, indicate, larger dose rate effects can be measured in large samples with 3D optical CT readout. This suggests cuvettes, 10 cm long with 2.2 cm inner diameter are not sufficient to reproduce 3D gel response. The data suggests that focus on maximizing dose sensitivity may result in formulations with measurable dose rate effects.

5. References

[1] Babic S *et al* 2009 *Phys. Med. Biol.* **54** 6791
[2] Vandecasteele J *et al* 2011 *Phys. Med. Biol.* **56** 6217
[3] Ebenezer S B *et al* 2013 *J. Phys.: Conf. Ser.* **444** 012027

Journal of Physics: Conference Series **2167** (2022) 012008 doi:10.1088/1742-6596/2167/1/012008

[4] Nasr A T *et al* 2015 *Phys. Med. Biol.* **60** 4665
[5] Lee S H *et al* 2014 *J Korean Physical Soc.* **64** 1063
[6] Jordan K *et al* 2017 *J. Phys.: Conf. Ser.* **847** 012009

Journal of Physics: Conference Series **2167** (2022) 012009 doi:10.1088/1742-6596/2167/1/012009

Simulated design optimization of a prototype solid tank optical CT scanner for 3D radiation dosimetry

A Ogilvy[1], S Collins[1], W Hare[1], M Hilts[2,1], T Tuokko[1], R Deardon[3] and A Jirasek[1]
[1]Department of Physics, University of British Columbia - Okanagan campus, Kelowna BC V1V 1V7, Canada
[2]Medical Physics, BC Cancer - Kelowna, Kelowna BC V1Y 5L3, Canada
[3]Department of Mathematics & Statistics and Faculty of Veterinary Medicine, University of Calgary, Calgary AB T2N 1N4, Canada

E-mail: andrew.jirasek@ubc.ca

Abstract. Optical computed tomography (CT) is one of the leading modalities for imaging gel dosimeters. There exist many prototype designs, as well as some commercial optical CT scanners that have showcased the value that gel dosimeters can provide to improve 3D dose verification for radiation treatments. However, due to factors including image accuracy, scan time, or demanding setup and maintenance there is currently no single scanner that has become a ubiquitous staple in a clinical setting. In this work, a prototype solid tank optical CT scanner is proposed that minimizes the need for a refractive index bath commonly found in optical CT systems. In addition to the design proposal, a ray-path simulator was created to optimize the design such that the solid tank geometry improves light collection across the detector array, maximizes the volume of the dosimeter scanned, and maximizes the dynamic range of the scanner.

1. Introduction

Gel dosimeters offer great potential as tools used to advance radiation treatment capabilities through three-dimensional dose verification [1]. By some criteria they are currently viewed as the only true 3D dosimetric tool [2]. A gel dosimeter is a composition of radio-sensitive chemicals that are fixed in place by a material matrix, most commonly gelatin, agarose, or plastic. The radio-sensitive chemicals respond in a measurable way to absorbed dose, and can then be read out at a later time [3]. Gel dosimeters can be imaged using MRI, x-ray CT, or optical CT [4]. The current state of gel dosimetry is quite open. Many varieties of gel dosimeters exist, and in optical CT, many gel dosimetry scanner designs. There is no obvious gold standard for a gel dosimeter or an imaging modality.

Optical CT is a high-resolution candidate for scanning gel dosimeters [5]. The imaging modality has good contrast, and a wide range of customizability for a relatively low-cost. Customization comes by way of optical instruments such as polarizing lenses, varied light sources, mirrors, and beam spreaders. However, optical CT relies on visible light which is subject to Snell's Law and the Fresnel Equations as it transmits a material interface. The resulting reflections increase noise via scatter, and refraction is a source of artifacts during image reconstruction, particularly filtered back projection.

Filtered back projection takes advantage of the reverse radon transform which assumes a direct path is taken from the signal source to the detector. However, in optical CT systems, when a photon transmits an interface of two media with differing refractive indices at a oblique angle, it the photon will refract. As a means to minimize refractive artifacts, nearly all optical CT designs contain some sort of index matching fluid contained in a bath to reduce refractive deviations [6, 7, 8, 9, 10]. Refractive baths typically have a significant volume of index matched fluid in them (1-15L). The fluid can evaporate or separate over long scanning periods, changing the remaining fluid's refractive index. If a matching fluid becomes inhomogeneous from separation, Schlieren patterns may appear in the viewing area naturally or as the result of motion in the fluid [11].

The goal of this research was to find a verifiably ideal design for a solid tank optical CT scanner. Such as design would maximize light collection, minimize artifacts, reduce the need for index matching, and have a uniform beam profile. An optical ray simulator was created using Matlab (MathWorks Natick, MA), was subsequently used to simulate hundreds of thousands of potential geometries with variations across five geometric variables. In this work, a specific scanner type was being tested which featured a solid acrylic "tank" allowing for a significant reduction in index matching fluid to between 5-15mL, and features the inclusion of iterative reconstruction techniques that account for refraction allowing for accurate image reconstructions regardless of refractive index mismatches. This work creates a framework for identifying ideal geometries for such a prototype fan-beam optical CT scanner, and optimizes for beam uniformity, spatial resolution, detector efficiency, and artifact minimization.

2. Materials and methods

The proposed fan-beam scanner design features a solid acrylic (PMMA) block with a hole bored through it. A gel dosimeter can then be placed through the bored hole, and moved using rotational and translational motors. An O-ring is fitted into the bore such that when pressed against a gel dosimeter it creates a seal. Within the gap between the bore and the dosimeter, a very small volume of refractive matching fluid is held in place by the O-ring. On the rear wall of the block is an array of uncollimated photodiode detectors, which have been their detecting surface pressed directly against the acrylic.

The primary tunable geometric properties that influence the quality of an optimal geometry were: the length of the block (x_{bl}), the position of the bore translated from the center of the block (x_{bc}), the distance from the focal point of the laser to front face of the block (x_{lp}), and two properties tied to the elliptical front face of the block, the semi-major axis length (x_{ma}) and eccentricity (x_{be}) (Fig. 1).

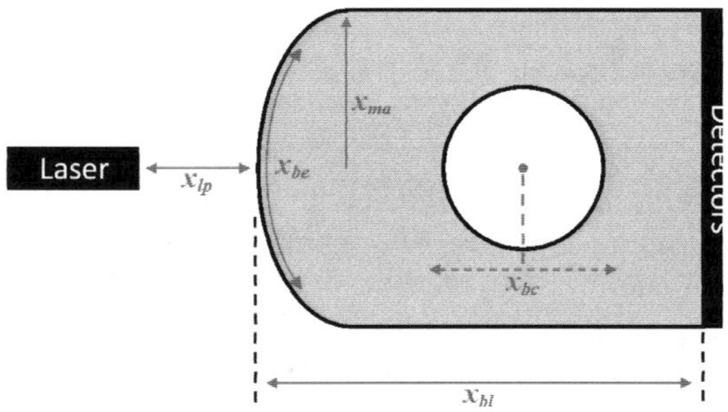

Figure 1: Shown is a schematic representation of the proposed solid tank optical CT scanner. Each of the five tunable parameters highlighted: block length (x_{bl}), bore position from the block center (x_{bc}), laser position from the block face (x_{lp}), semi-major axis length (x_{ma}) and block eccentricity (x_{be}).

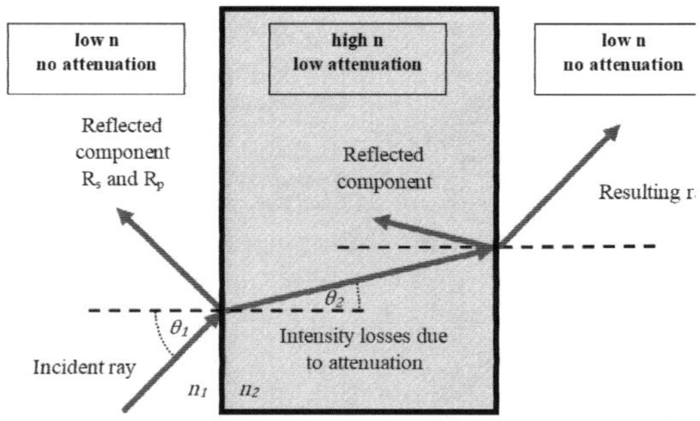

Figure 2: An example showing how ray path and intensity are simulated through a geometry. At refractive interfaces the ray-path is altered (Snell's). An incident ray suffers an intensity loss from reflection upon the entry and exit from an object (Fresnel) with a different refractive index, as well as suffering intensity losses from attenuation (Beer-Lamber).

Journal of Physics: Conference Series **2167** (2022) 012009 doi:10.1088/1742-6596/2167/1/012009

The ray-path simulator that was constructed has full customizability including: the five tunable parameters, laser polarity, material refractive index, attenuation coefficients, laser fan angle, etc. During each simulation, 400,000 rays are generated and uniformly distributed in a 60° fan, approximating the optical fan-beam created by our laser. In this research the virtual gel was comparable to a FlexyDos3D dosimeter with a refractive index of 1.4225, and a uniform linear attenuation of 0.166 cm^{-1} [12].

When a ray a material interface, the refraction induced changes to ray path is calculated using Snell's Law. The relative intensity of each ray is tracked as it passes through the system, measuring losses from reflection using the reflectance Fresnel equations, as well as intensity losses from attenuating objects which can be placed in the system (Fig. 2).

The purpose of simulating many geometries is to find a best geometry. A best geometry would maximize spatial resolution, minimize spatial artifacts, and maximize dynamic range of detector profiles. In order to score geometries three relevant quantities were identified: effective radius (Fig. 3), magnification (Fig. 4), and beam uniformity (Fig. 5). Effective radius evaluates how much of the bore area is imaged, and highlights any reductions in the field-of-view (FOV) is due to ray refractions. The magnification quantifies the fraction of the available detectors that are used to image the gel volume. Beam uniformity is the root-mean-square error (RMSE) of each simulated signal profile. Each of the three properties was weighted equally and scored from 0 to 1. The sum of the scores for each property was the final metric used to identify ideal geometries (Fig. 6).

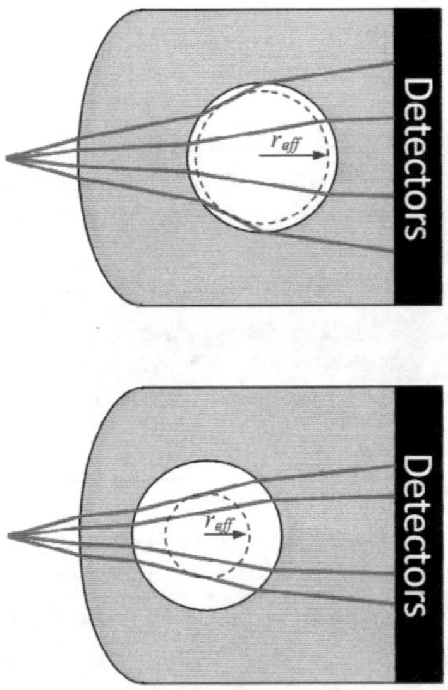

Figure 3: Schematic showing ray paths that result in different effective radii (r_{eff}). Top panel: ray paths resulting in high r_{eff}. Bottom panel: ray paths resulting in low r_{eff}.

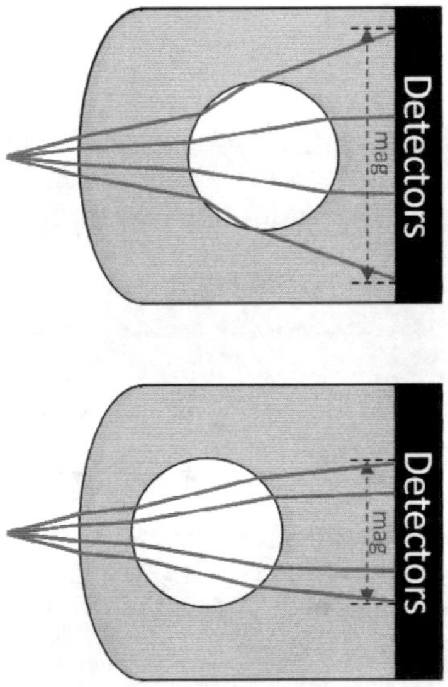

Figure 4: Schematic showing ray paths that result in varied magnifications. Top panel: ray paths leading to high magnification. Bottom panel: ray paths leading to low magnification.

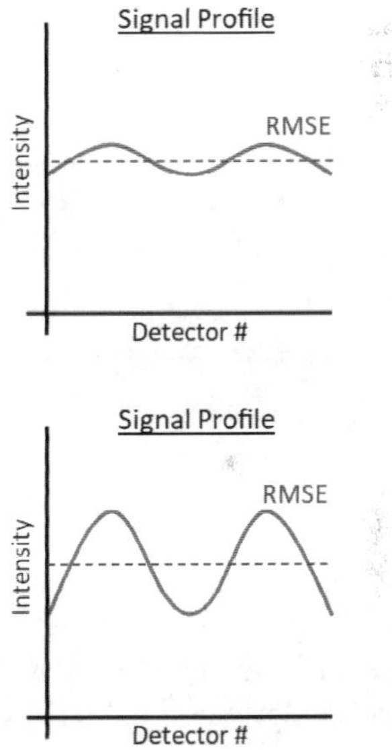

Figure 5: A sample diagram showing intensity profiles that would result in different beam uniformity (RMSE) values. Top panel: a signal profile with good beam uniformity. Bottom panel: a signal profile with poor beam uniformity.

3. Results and discussion

During simulation, virtual detectors are used as bins to collect rays refracted through the solid tank. The number of rays passed through the system was tuned in order to ensure that a statistically significant number of rays are collected in each detector element. Increasing the number of rays results in a more accurate profile, but increases the time required for each simulation, and 400,000 rays was found to be a balance between computation time, and profile smoothness.

Each simulation dataset returns a 5 dimensional matrix of scores corresponding to the permutations of the 5 variables: x_{bl}, x_{bc}, x_{lp}, x_{ma}, and x_{be}. During simulation a challenge was sampling resolution. The nature of 5D computation means that to go from 10 points of resolution to 11 would be a 61% increase in the number simulations for a 9% increase in resolution. Ultimately, trial and error was the method used to find regions of high-scoring geometries. Once one such area was found, another "zoomed-in" simulation was performed to get a higher sampling resolution in the region-of-interest.

The ideal values for the 5 variables were found to be: $x_{bl} = 290$mm, $x_{bc} = 5$mm, $x_{lp} = 41$mm, $x_{ma} = 61$mm, and $x_{be} = 0$. The beam profile of this geometry can be seen in Fig. 7, and to show how quickly beam uniformity degrades Fig. 8 shows a profile with $x_{bc} = 18$mm.

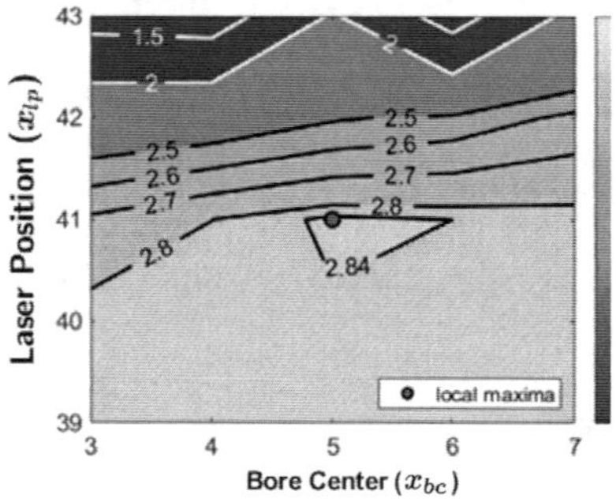

Figure 6: A Laser position (x_{lp}) vs. Bore Center (x_{bc}) 2D contour of 5D data which shows a local maxima. Fixed variables are: x_{bl}, x_{ma}, and x_{be}.

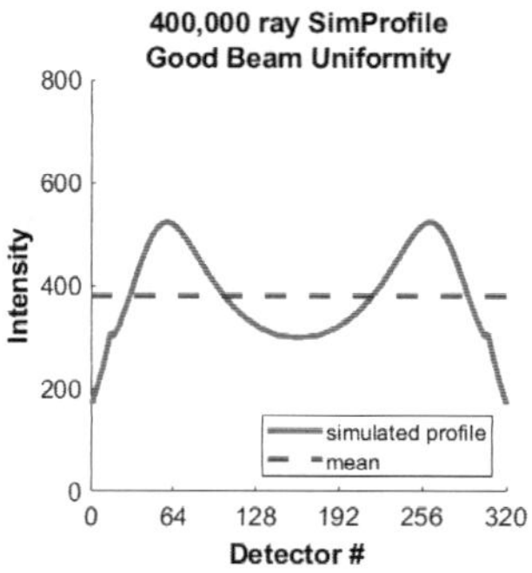

Figure 7: A simulated profile with decent beam uniformity ($x_{bc} = 5$mm).

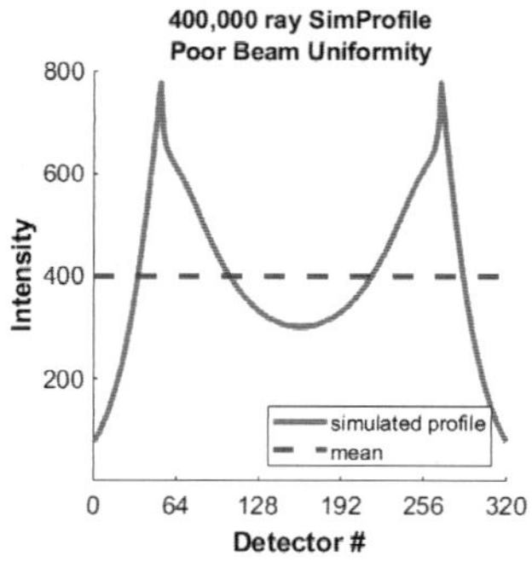

Figure 8: A simulated profile with poor beam uniformity ($x_{bc} = 18$mm).

4. Conclusion

Using a ray path simulator, an ideal design for the proposed solid tank fan-beam optical CT scanner was found. Hundreds of thousands of geometries were simulated by varying 5 variables corresponding to the block geometry. The simulated geometries were compared by scoring three properties of interest: effective radius, magnification, and beam uniformity. The ideal values for each of the variables were found to be: $x_{bl} = 290$mm, $x_{bc} = 5$mm, $x_{lp} = 41$mm, $x_{ma} = 61$mm, and $x_{be} = 0$ for this specific design. This simulator can be used in the future to find the optimal design of a new block that would be used to test gels other than FlexyDos3D.

5. References

[1] Jackson J *et al* 2015 *Phys. Med. Biol.* **60** 2217-30
[2] Schreiner L J 2015 *J. Phys. Conf. Ser.* **573** 012003
[3] Hoecker F E *et al* 1958 *Int. J. of Appl. Radiation and Isotopes* **31** 35
[4] Baldock C *et al* 2010 *Phys. Med. Biol.* **55** R1-63
[5] Oldham M 2006 *J. Phys. Conf. Ser.* **56** 58-71
[6] Gore J C *et al* 1996 *Phys. Med. Biol.* **41** 2695
[7] Wolodzko J G *et al* 1999 *Med. Phys.* **26** 2508-13
[8] Krstajić N *et al* 2007 *Phys. Med. Biol.* **52** 3693-13
[9] Xu Y *et al* 2004 *Med. Phys.* **31** 3024- 33
[10] Campbell W G *et al* 2013 *Med. Phys.* **40** 061712
[11] Settles G S 2001 *Schlieren and Shadowgraph Techniques Visualizing Phenomena in Transparent Media.* SpringerLink eBooks
[12] Deene Y D *et al* 2015 *Phys. Med. Biol.* **60** 1543-63

Rapid manufacture of patient-specific, elastomeric, three-dimensional dosimeters using the FlexyDos3D dosimeter

M J Wheatley[1], and Y De Deene[1,2]
[1]School of Engineering, Macquarie University, Sydney, Australia
[2]Radiology Department, Nepean Blue Mountains Local Health District, Sydney, Australia

E-mail: Morgan.Wheatley@hdr.mq.edu.au

Abstract. Patient-specific dosimeters are ideal for quality assurance of radiotherapy treatments. Such dosimeters would ideally need to be manufactured in a short amount of time so that the time between initially imaging the patient, treating the patient-specific dosimeter, reading out the dosimeter, and then finally treating the patient, is minimised. 3D printing allows for rapid manufacture of complex 3D objects. The FlexyDos3D dosimeter is a silicone-based material that has enough strength for the dosimeter to hold its shape without a mould or other structure supporting it. A custom 3D printer has been built for making dosimeters from the FlexyDos3D material. Testing of the 3D printer is ongoing.

1. Introduction

Quality assurance of radiotherapy treatments typically rely on point detectors, such as ionisation chambers, 2D films, or 3D dosimeters. 3D dosimeters allow for the dose to be measured throughout the dosimeter in three dimensions [1]. However, most 3D detectors are cast into moulds to provide the shape of the dosimeter required and many of these moulds are of simple shapes, such as cylindrical moulds [2]. Some groups have investigated the use of 3D printers to manufacture the moulds used to hold these 3D dosimeters [3-5]. However, the material printed is not the dosimeter itself and is instead used to mimic the geometry and CT or MRI properties of a patient whilst being filled with a dosimeter. 3D printing of the dosimeter itself would allow for more complex objects to be made that could allow for hollow chambers in the dosimeter, like those in the heart or lungs.

For a dosimeter to be rapidly constructed and match a patient's anatomy without the use of a mould or container holding the dosimeter, a dosimeter with high strength is required such that the material can stand up under its own weight. The FlexyDos3D three-dimensional dosimeter [6] consists of a silicone matrix which gives the dosimeter strength, and leucomalachite-green (LMG) and chloroform to cause the optical change in the dosimeter that can be correlated to dose.

Silicone, when at room-temperature, takes about 48 hours to solidify but will solidify faster with heat. Therefore, to rapidly set the material, high temperatures are required. The silicone is also a thermosetting plastic meaning that once it sets, it will not melt even with high temperatures, unlike other Fricke gel or polymer gel dosimeters.

A custom-built 3D printer has been constructed that allows for the 3D printing of the FlexyDos3D dosimeter. The printer is still undergoing testing and only simple geometric shapes have been created so far. The dosimeter still shows a dose response when irradiated after printing.

2. Methods and Materials

2.1 Dosimeter recipe

The FlexyDos3D dosimeter consists of a silicone elastomer (Sylgard® 184, Merck) with 4 % w/w curing agent and 93 % w/w base material of the silicone, 3 % w/w chloroform, and 0.03 % w/w LMG (the sum of the percentages do add up to 100.03 % w/w but the 0.03 % w/w difference due to the LMG is seen as negligible). The curing agent and base material are mixed thoroughly first. The LMG is first mixed into the chloroform before that mixture then being stirred into the curing agent and base mixture. Many small air bubbles will be in the mixture due to the stirring. These air bubbles will eventually rise and settle out of the mixture, but a vacuum chamber can be used to speed up this process. Care should be taken when using vacuum to remove the bubbles as the mixture can grow more than three times its volume when the air bubbles in the mixture expand in vacuum. Care should be taken to avoid exposing the dosimeter to bright light and UV sources, or high temperatures to avoid causing the dosimeter to react prematurely.

2.2 3D printer construction

To rapidly cure the silicone dosimeter, high temperatures are required. An oven enclosing the print volume is used to heat the dosimeter once it is extruded out of a nozzle. The oven is heated to a temperature of 160 °C. To ensure that the dosimeter material is not curing inside the tubing whilst it is being pumped into the oven and out of the nozzle, another tube with circulating, room-temperature water surrounds the tubing with the dosimeter material in it. The water is kept at, or just above, room-temperature (22 °C) to avoid condensation on the cooled components which could damage electronics or cause water to get inside the oven enclosure.

The motors used to move the print nozzle are only rated to withstand temperatures up to 60 °C. As such, they cannot be located within the oven enclosure. A delta-style printer configuration was used whereby three pairs of motors are used with all the motors shaft axes oriented in the same direction above the printer (Fig 1). Each pair of motors has an individual moving platform that can be moved up and down a threaded shaft by their respective motor. Each platform then has a set of rods connected to it and each of these rods other ends connect to a central platform that houses the print nozzle. By moving each of these platforms interdependently, the nozzle can be moved in the three-dimensions within the print volume.

As the motors are attached to the oven by their metal shafts, heat can conduct through the shaft into the motor. To keep the motors cool, the same tubing that contains the cooling liquid pumped around the dosimeter material being extruded into the oven is also routed through some custom-built heatsinks which are bolted to the motors. These are used to remove heat from the motors and keep them beneath 60 °C.

A custom printed circuit board (PCB) was made to house the electronics and the microcontroller that controls the printer. The open-source Repetier software (Hot-World GmbH & Co. KG) was used both for the firmware on the microcontroller that controls each of the printer's operations, and the software on the computer that allows the commands to be sent from the computer to the microcontroller. The software commands are generated from the 3D model on the computer to automate the printing.

Journal of Physics: Conference Series **2167** (2022) 012010 doi:10.1088/1742-6596/2167/1/012010

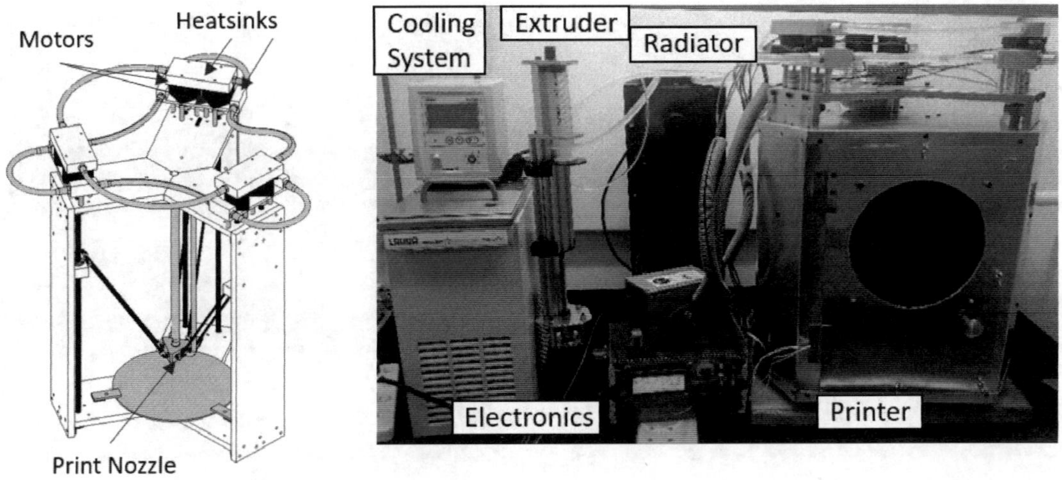

(a) (b)

Figure 1. A computer 3D model is shown of the printer without the surrounding oven enclosure and the fan to show the printer (a). The motors can be seen mounted above the printer (with the heat-sinks on-top and below) in the delta-configuration. An image of the printer with the electronics control box (bottom), cooling system (far left), extrusion system (syringe and motor assembly to the right of the cooling system), radiator and fans (rear centre), and the printer (right) (b). There is a glass viewing window seen on the door to the printer which allows for viewing of the print job during printing.

3. Results and Discussion

The printer is currently being tested to optimise print speeds, temperatures, nozzle diameters, and layer heights. Basic shapes have been made to test the system, such as hollow cylinders, hollow cubes, and solid cubes. Measurements are being performed on the geometric accuracy of these simple shapes as they are easy to measure and are dependent upon the above parameters.

Photos of a printed hollow cube (made of just silicone for ease of testing) and a solid cylinder and U-shaped object (both made of the FlexyDos3D material) are both shown below (Fig 2). As shown in the pictures, the printed objects do have the rough dimensions of the intended object. However, the side surfaces of the objects are irregular due to the poor accuracy of the current printer mechanics. The U-shaped object was irradiated with a UVC source and a stencil to show the logo of the IC3DDose conference logo to demonstrate the material still reacts to ionising radiation after printing. Further testing is still required to determine the dose response of the dosimeter after printing.

The current material extrusion rate is about 0.3 mL per minute which is on par with many commercial 3D printers. The shown printed objects each took roughly 20 minutes to print except for the U-shaped object which took roughly 40 minutes to print.

Once the mechanical accuracy of the printer is improved, 3D images of patients can be taken, and sections of the patient can be made with the printer.

Journal of Physics: Conference Series **2167** (2022) 012010 doi:10.1088/1742-6596/2167/1/012010

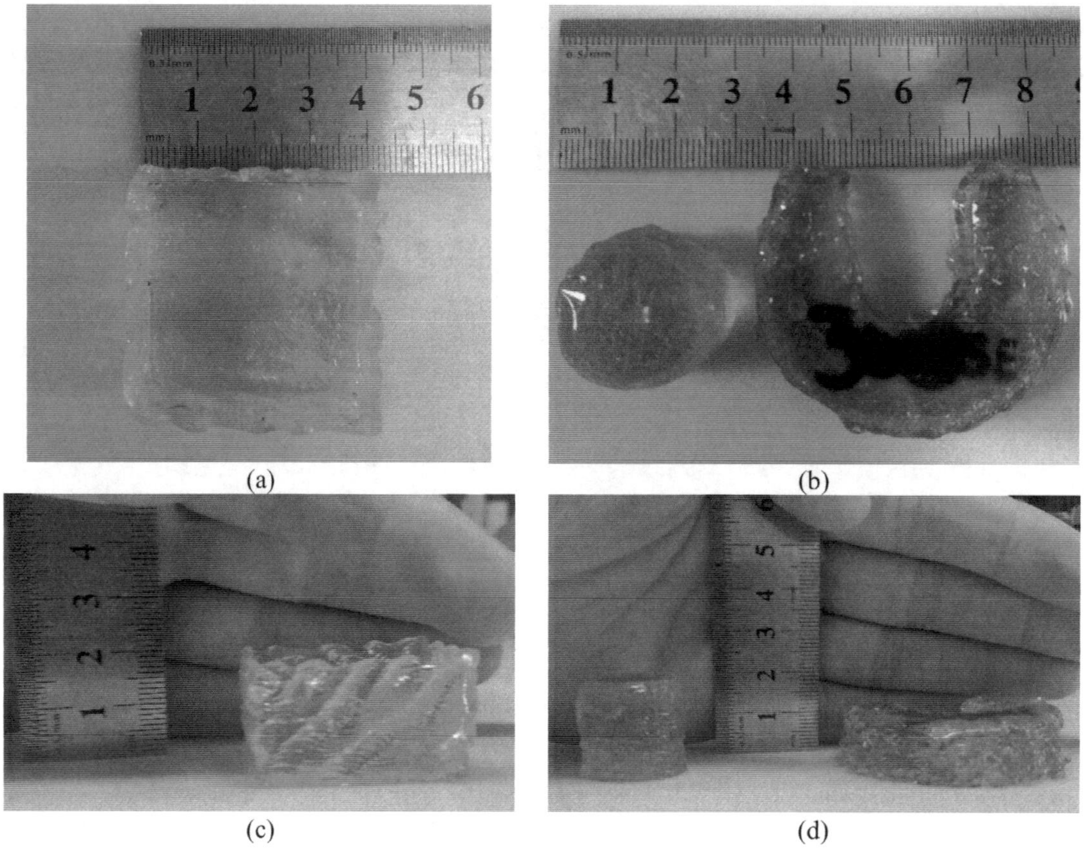

Figure 2. Photos of some test objects printed by the printer. A hollow cube printed with silicone with dimensions of 4 cm x 4 cm x 2 cm with a wall thickness of 1 mm (a, c). However, due to inaccuracies with the current mechanical setup, the thickness of the walls can vary between 1 mm and 2 mm, as seen by the pattern in the sides of the print (c). A solid cylinder of dimensions 2 cm diameter x 2 cm height and a U-shaped object with dimensions of 5 cm diameter x 1 cm high with a removed inner section of 2 cm width is shown (b, d). A stencil and UVC light was used to irradiate the IC3DDose conference logo into the U-shaped object, demonstrating the printed material is still able to react to ionising radiation.

4. Conclusion

Creating simple-shaped dosimeters for radiotherapy is possible with the FlexyDos3D material and the accompanying 3D printer. Currently, no quantitative irradiations have been performed on the printed objects and the printed jobs have been limited to small, simple shapes. However, it is believed that larger-scale phantoms, the size of the heart or brain, will be able to be constructed once the accuracy of the printed objects can be improved.

The printer can also print silicone material and so has an application to print 3D moulds that can hold other dosimeter material, such as Fricke gels or polymer gels, and form a flexible mould. Deformations can then be performed upon these moulds and the dosimeter inside them.

To minimise print times for larger dosimeters, the external shape of the dosimeter could be printed with the internal structure remaining mostly hollow. The dosimeter could then be filled with more of the dosimeter material.

5. Acknowledgements
The authors would like to thank David Baer, Elton Button, and Walther Adendorff and the MQ METS staff for their consultation and provision of equipment and material in the design and construction of the system.

6. References
[1] Baldock C *et al* 2010 *Phys. Med. Biol.* **5** R1-R63
[2] Costa F *et al* 2018 *Phys. Med. Biol.* **63** 05NT01
[3] Oh D *et al* 2017 *Scientific Reports* **7** 40922
[4] Ehler E D, Barney B M, Higgins P D *et al* 2014 *Phys. Med. Biol.* **59** 5763-73
[5] Mayer R, Liacouras P, Thomas A *et al* 2015 *Review of Scientific Instruments* **86** 074301
[6] De Deene Y, Skyt P S, Hill R *et al* 2015 *Phys. Med. Biol.* **60** 1543-63

Journal of Physics: Conference Series **2167** (2022) 012011 doi:10.1088/1742-6596/2167/1/012011

On the feasibility of using an optical fiber Bragg grating array for multi-point dose measurements in radiation therapy

Marie-Anne Lebel-Cormier[1,2,3], Tommy Boilard[1,4], Martin Bernier[1,4] and Luc Beaulieu[1,2,3]

[1]Département de physique, de génie physique et d'optique, Université Laval, Québec, Canada

[2]Centre de recherche sur le cancer, Université Laval, Québec, Canada

[3]CHU de Québec - Université Laval and CRCHU de Québec, Canada

[4] Centre d'optique, photonique et laser (COPL), Université Laval, Québec, Canada

E-mail: marie-anne.lebel-cormier.1@ulaval.ca, Luc.Beaulieu@phy.ulaval.ca

Abstract. Fiber Bragg gratings (FBGs) have proven to be a valuable dosimeter in nuclear environment where radiation doses reach up to a hundred of kiloGray (kGy). Multiple FBGs can be written in a single fiber to allow multi-point detection which would prove very useful for radiotherapy dosimetry. The purpose here is to adapt this already existing technology to provide a novel dosimeter for radiotherapy measurements. The proposed real-time dosimeter consists of twenty 4 mm-long FBGs, equally distributed over 20 cm. FBGs are written through the coating of a standard polyimide-coated silica fiber with the phase-mask technique and femtosecond pulses. The wavelength dependant variation of each FBG is recorded at 1 kHz with a commercially available interrogator. The use of gamma radiation (clinical radiotherapy accelerator) induces a linear shift (0.070 $\pm$ 0.006 pm/Gy) of the FBG's reflected wavelength, which is independent of the dose rates (2.8-11.6 Gy/min) and the energy (6-23 MV). A statistical error of 0.03 pm is obtained on data points therefore limiting the detectable dose to 0.4 Gy. A dose profile of 6 and 23 MV radiotherapy accelerator is also measured. The presented FBGs dosimeter allows for real-time dose measurement in 2D and the small size of its detector makes it a versatile tool. The length and spacing of FBGs can be easily modified to increase both the spatial resolution and the amount of dose point.

1. Introduction

Fiber Bragg gratings (FBGs) are now common and versatile sensors used in a wide range of applications in the industries to measure pressure, stress or temperature. They can be used to detect leaks on a pipeline, but also measure the deformation of the wings of an airplane during flights [1-2]. A FBG is a periodic or quasi-periodic modification of an optical fiber core refractive index relative to its propagation axis. When light propagates through this periodic refractive index variation, it produces a reflection of a narrow bandwidth of light centered around the Bragg wavelength. A variation of temperature or strain applied to the FBGs sensors produce a reflected wavelength shift. Thus, the FBGs sensors determine the quantity of interest (pressure, stress, temperature, etc.) by measuring the reflected wavelength shift. FBGs sensor has also proven to be a useful dosimeter in nuclear environment where dose reach up to a hundred kGy [3]. In these conditions, radiation produces damages to the fiber glass, which modifies the fiber refractive index according to the Kramers-Kronig relation [4]. This results in a reflected wavelength shift that can be measured with a dedicated interrogator and correlated to the emitted radiation. Since many FBGs can be written in a single fiber, this dosimeter allows, by design, multi-point detection. A single point prototype relying on post-irradiation thermo-optic properties of FBGs, therefore preventing real-time use, was proposed with a reported minumum detection of 0.160 Gy [5].

Hence, the purpose here is to adapt this technology to provide an innovative real-time dosimeter for radiotherapy measurements. For this proof-of-concept, an optical 20 points real-time dosimeter, using fiber Bragg gratings, is proposed for clinical radiation dosimetry measurements. Preliminary characterization of this dosimeter response to dose and dose rates is presented in this paper as well as dose profile measurements.

2. Methods and materials

2.1. FBGs inscription
The FBGs are written by focusing 35 fs pulses at 800 nm (Coherent, Astrella) through the coating of a standard 125 μm polyimide-coated fiber (OFS BF06160-02) by a short focal acylindrical lens [6]. A custom e-beam phase mask with 20 uniform periods ranging from 1045 nm to 1091 nm was used to write an array of 20 different 4-mm long FBGs along the fiber length. The FBGs are spectrally distributed from 1516 nm to 1581 nm, with a separation of 3.4 nm between each peak.

2.2. Irradiation conditions
All irradiations were performed on a CLINACiX. Shifts in wavelength due to irradiation of the FBGs by a 6 and 23 MV beam were recorded for a dose up to 20 Gy. A 6 Gy/min dose rate and 10 X 10 cm field size were used unless otherwise noted.

2.3. Experimental set-up
The detector is made of twenty (20) 4mm-long FBGs, equally distributed over 20 cm. The detector is fixed to a PMMA sheet to minimize environmental factor variation such as pressure applied on the detector. The plastic sheet also ensures that the detector is tightly surrounded by diffusing material. The wavelength shift of each FBG is recorded at 1 kHz with a commercially available interrogator (si155, Micron Optics) and its sensing analysis software (ENLIGHT). The data points are averaged to have one measure per second. As shown in figure 1, a diffusing material (solid water) is used and the detector is placed at d_{max} (1.5 cm for 6 MV and 3.5 cm for 23 MV).

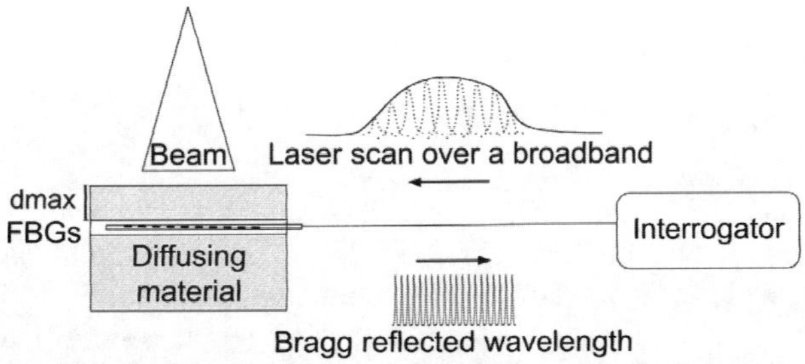

Figure 1. Experimental set-up. As the laser beam spreads into the FBGs, the reflected wavelength is measured with the interrogator.

Journal of Physics: Conference Series **2167** (2022) 012011 doi:10.1088/1742-6596/2167/1/012011

3. Results and Discussion

3.1. Dose and dose rates dependence

A linear reflected wavelength shift is obtained upon irradiations. Using 21 dose measurements (3 trials of 7 FBGs), the mean signal slope is 0.070 pm/Gy and the standard deviation is 0.006 pm/Gy. Figure 2a shows the reflected wavelength shift of 1 grating produced by a 20 Gy irradiation. The standard error on data points is 0.03 pm, which, using the response to dose slopes, leads to a detection limit of this early prototype of 0.4 Gy. This detection limit is far from the 0.01 Gy dose resolution required for radiotherapy applications [7], but we are currently working on material optimization to make the dosimeter more sensitive to radiation which will increase the detection limit. It should also be noted that a higher sampling frequency would decrease the statistical error therefore resulting in an increased detection limit.

The detector is dose rate independent (tested between 2.8 – 11.6 Gy/min) as shown in Figure 2b. The presented values are averaged from seven measurements, except for the 6 Gy/min dose rate for which we averaged the Bragg reflected wavelength shift for twenty-one irradiations. The error bar corresponds to one (1) standard deviation.

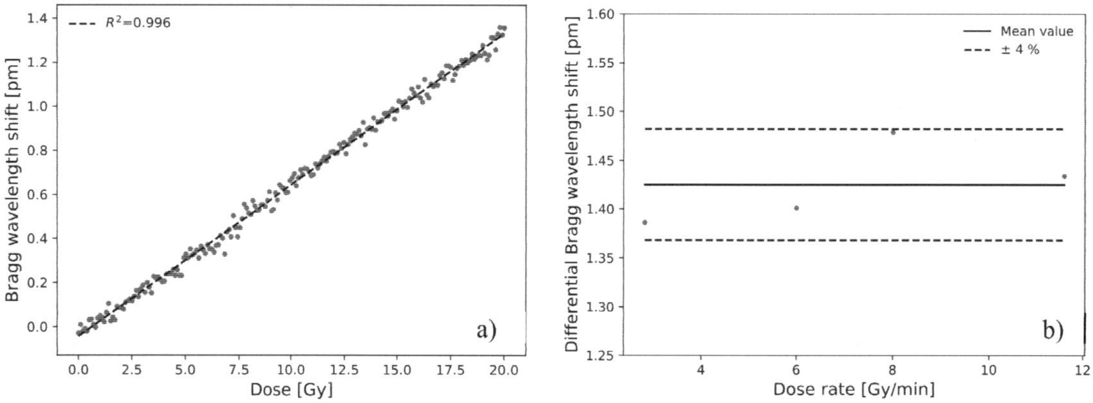

Figure 2. a) Bragg reflected wavelength shift in terms of dose and b) differential Bragg reflected wavelength produced by a 20 Gy irradiation (6 MV beam).

3.2. Dose profiles

A dose profile of 6 MV and 23 MV radiotherapy accelerator 10 x 10 cm beam is measured with a mean error of 6.3 ± 0.7 % and 7 ± 1 % on the central region (Figure 3). The signal variation can be explained by the fact that Bragg wavelengths are affected by temperature variation. We applied a temperature-based correction to minimize the impact of temperature variation on the detector response, but the method still needs improvements. The mean shift over 21 measurements (3 trials of the 7 FBGs in the central region) is 1.4 ± 0.1 pm for the 6MV beam and 1.4 ± 0.1 pm for the 23MV beam. Thereby, we conclude that the presented submillimeter-size detector is also energy independent between 6MV and 23 MV beams.

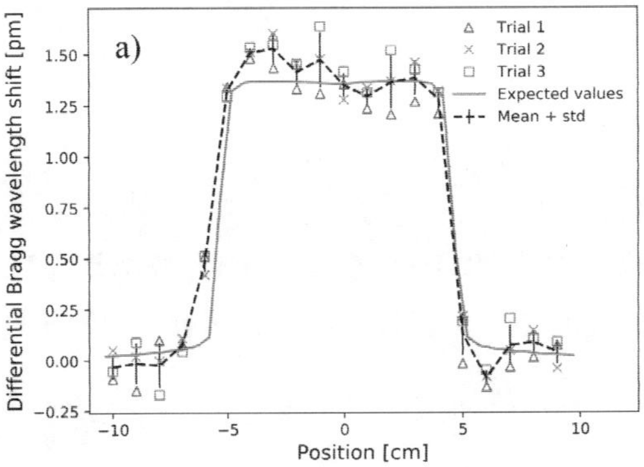

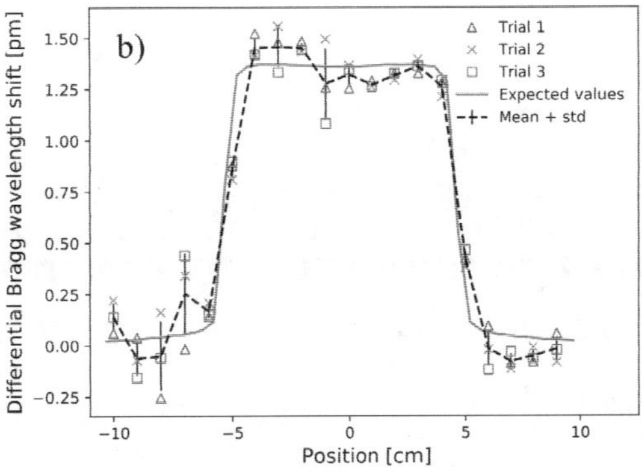

Figure 3. Dose profile for a 20 Gy irradiation using a) 6 MV and b) 23 MV beam. The mean and standard deviation of the three trials are presented as well as the expected values measured with Pinnacle (TPS).

4. Conclusion

We propose a novel FBG-based dosimeter that supports real-time dose measurements in 2D. The submillimeter diameter (155 microns diameter) of the detector makes it suitable for *in vivo* dosimetry, as well as regular and small field measurements. It allows, by design, multi-point dose measurements with a spatial resolution of 4 mm, which make dose profile measurements possible. Moreover, it has a linear response to dose while being independent to dose rate and energy. It is also important to mention that length of FBGs and spacing can be customized and hence, the spatial resolution and the amount of dose points can be increased significantly, up to 50 with the current assembly. We now plan on developing a prototype with suitable spatial resolution for small field dosimetry.

5. Acknowledgements

This work was supported by the Natural Sciences and Engineering Research Council of Canada (NSERC) Discovery grants RGPIN #2019-05038 and Natural Sciences and Engineering Research Council of Canada (CG112389).

6. References

[1] Joe H E, Huitaek Y, Jo S H *et al* 2018 *Int. J. of Precis. Eng. and Manuf.-Green Tech.* **5** 173-91
[2] Campanella C E, Antonello C, Campanella C *et al* 2018 *Sensors* **18** 3115
[3] Berghmans F and Gusarov A 2011 *Fiber Bragg Grating Sensors: Recent Advancements, Industrial Applications and Market Exploitation* ed Cusano A, Cutolo A, and Albert J (Bentham Science Publishers) 233
[4] Girard S *et al* 2019 *Reviews in Physics* **4** 100032
[5] Girard S *et al* 2018 *J. Opt.* **20** 093001
[6] Habel J, Boilard T, Frenière J S *et al* 2017 *Sensors* **17** 2519
[7] Rink A and Jaffray D A 2016 *Scintillation Dosimetry* ed Beddar S and Beaulieu L (New York : CRC Press) 295

Adaptive radiotherapy dosimetry in a challenging geometry: A model gas-filled tissue expander in a helical TomoTherapy beam

T. Kairn[1,2], M. Lathouras[1], S. R. Sylvander[1], J. V. Trapp[2] and S. B. Crowe[1,2,3]
[1]Cancer Care Services, Royal Brisbane and Women's Hospital, Herston Qld 4029, Australia
[2]Science and Engineerig Faculty, Queensland University of Technology, Brisbane Qld 4000, Australia
[3]Herston Biofabrication Institute, Herston Qld 4029, Australia

Email: t.kairn@gmail.com

Abstract. Helical TomoTherapy treatment and delivery systems (Accuray Inc, Sunnyvale, USA) allow off-line adaptation of radiotherapy treatments, with dose calculations that use MV computed tomography (CT) data acquired at treatment. This study aimed to assess the potential dosimetric effects of a gas-filled temporary tissue expander (TTE) on the accuracy of breast radiotherapy dose calculations from both the TomoTherapy treatment planning system (TPS), which uses kV CT data, and the TomoTherapy adaptive radiotherapy (ART) system, which uses MV CT data. A TomoTherapy treatment plan was created and delivered to a 3D-printed rectilinear model of a breast with implanted gas-filled TTE, including a stainless steel CO_2 container, and film measurements of the delivered dose were compared against dose calculations from the TPS and ART systems. The film measurements showed that the TomoTherapy TPS provided comparatively accurate dose calculations in the ~550 cm^2 volume of air that modelled the gas filling of the TTE and within the surrounding tissue-equivalent materials, except in regions where the beam was transmitted through the stainless steel CO_2 container, possibly due to the volume of stainless steel being over-estimated in the kV CT images that were used to generate the treatment plan. The ART system provided more accurate dose calculations than the TPS in regions affected by the stainless steel container, but also over-estimated the dose in the air within the TTE. These results suggest that the TomoTherapy TPS and ART systems could be used to produce reliable dose calculations of breast treatments in the presence of gas-filled TTEs, if kV CT imaging options are chosen to avoid artefacts and minimise the need for density over-rides and if treatment targets that include only clinically relevant tissues, and exclude all TTE components, are used to evaluate and compare the doses calculated by both systems.

1. Introduction

The Accuray Hi-Art helical TomoTherapy (HT) system (Accuray Inc, Sunnyvale, USA) allows off-line adaptive radiotherapy, by re-calculating treatment doses on patient MV computed tomography (CT) images acquired at treatment delivery and comparing the results against the planned treatment doses calculated initially using planning kV CT images [1-3]. This method allows efficient identification of and adaptation to the dosimetric effects of changes in patient contour, target size or treatment setup [3].

A secondary advantage of the use of MV CT data to verify dose distributions planned using kV CT data is the low susceptibility of MV CT images to reconstruction artefacts from high-density implants [3]. However, MV images also provide reduced bone-tissue contrast compared to kV images [4] and MV CT are often optimized to limit patient dose by allowing increased noise [2]. This can mean that the intrinsic differences between kV treatment planning images and MV treatment verification images, in terms of contrast, noise and artefacts, can result in ambiguity and confusion when interpreting the results of adaptive radiotherapy dose comparisons: When results differ, especially in the presence of density heterogeneities, which system should be trusted?

Due to the precipitous density heterogeneities that result from their particular structure and function, gas-filled temporary tissue expanders (TTEs) were selected for investigation in this study, as a useful worst-case-scenario. TTEs are used in two-stage breast reconstructions, after mastectomies for breast cancer [5,6]. Breast TTEs are usually filled with saline via a magnetic port [5] but can be filled with CO_2 via a metallic gas canister located within the implant [6]. Gas-filled TTEs are therefore expected to detrimentally affect the accuracy of radiotherapy treatment dose calculations due to the presence of a high-density (mostly stainless steel) component, within a large low-density (gas) volume, covered by a thin layer of skin and residual breast tissue. Previous studies have shown how such implants can affect static photon beams [7,8] and electron beams [9], but effects on TomoTherapy dose calculations and MV CT-based adaptive radiotherapy dose calculations have not been reported.

This study used multi-plane film measurements to evaluate the accuracy of dose calculations provided by both the TomoTherapy "Hi-Art" treatment planning system (TPS) and the TomoTherapy "Planned Adaptive" adaptive radiotherapy (ART) dose re-calculation and evaluation system. A model gas-filled TTE was developed, to provide a rectilinear but realistically heterogeneous test object, for evaluating the performance of the TomoTherapy TPS and the ART systems under geometrically and dosimetrically challenging conditions that were similar to the treatment of a breast radiotherapy patient with a gas-filled TTE.

2. Method and materials

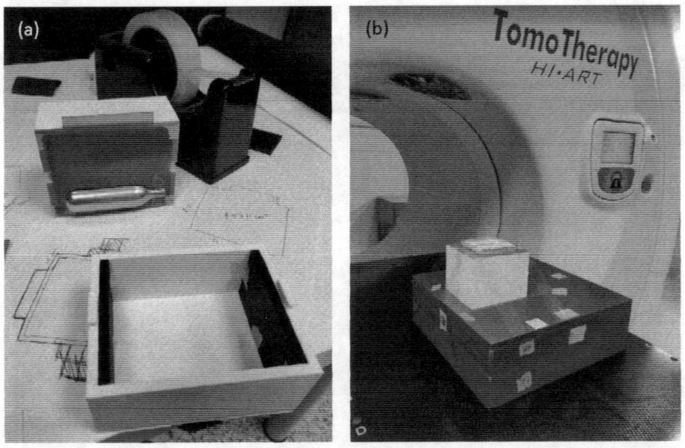

Figure 1. Photographs of phantom, showing 3D-printed breast containing air volume, CO_2 container and film (a), and whole phantom assembly, with Virtual Water, film and bolus (b).

Previous studies of the effects of gas-filled TTEs on the dose from breast radiotherapy treatments have compared TPS dose calculations for static tangent beams against point dose measurements on humanoid phantoms with sample TTEs [7,8] or on the skin of a patient with a TTE in situ [10]. To provide additional information on the dose in and around gas-filled TTEs, while providing two-dimensional dose measurements appropriate for evaluating a modulated radiotherapy beam [11], a simple phantom geometry was constructed using an additive deposition 3D-printing technique.

The 3D-printed components of the phantom were designed using TinkerCAD software (Autodesk Inc, San Rafael, USA), to house a stainless steel CO_2 container and a ~550 cm^2 volume of air, approximately modelling a 0.5 – 1.0 cm thick layer of tissue covering a gas-filled TTE (see figure 1(a)). This simulated "breast tissue" was designed to be rectilinear and printed in two halves, to allow film measurements within the model implant that could be unambiguously compared against dose calculations provided by the TPS and ART systems. An Ultimaker 2 Extended+ 3D printer was used to fabricate the model, using PLA filament, with a 90% in-fill density. This material and density were chosen due to their established radiological tissue-equivalence [12].

The 3D-printed breast model, containing the CO_2 container and pieces of Gafchromic EBT3 radiochromic film, was positioned on top of a 30 × 30 × 10 cm^3 slab of Virtual Water (Standard Imaging Inc, Middleton, USA), with a small piece of bolus placed on top of the breast at the location where a mastectomy scar would occur, duplicating local clinical procedure. This setup (shown in figure 1(b)) was used during acquisition of kV CT image data (used for treatment planning), as well as during acquisition of pre-treatment MV CT image data (used for treatment setup verification and ART dose calculation) and during treatment delivery. More than 20 h after treatment delivery, the film was scanned and results were converted to dose [11] before being compared with dose calculations from the TomoTherapy TPS and ART systems.

3. Results and discussion

The kV CT of the phantom was observed to be affected by image reconstruction artefacts, arising from the presence of the metallic CO_2 container within the model breast. The affected regions of the kV CT were therefore contoured and over-ridden with realistic density values, for air, water and stainless steel, before treatment planning. The MV CT of the phantom, acquired at treatment, was observed to be noisier than the kV CT but unaffected by artefacts, so no overrides were applied before dose calculation by the ART system.

Figures 2(a)-(d) show the results of film measurements at various locations within the phantom and demonstrate the effects of the model gas-filled TTE on the accuracy of dose calculations provided by the TPS and ART systems.

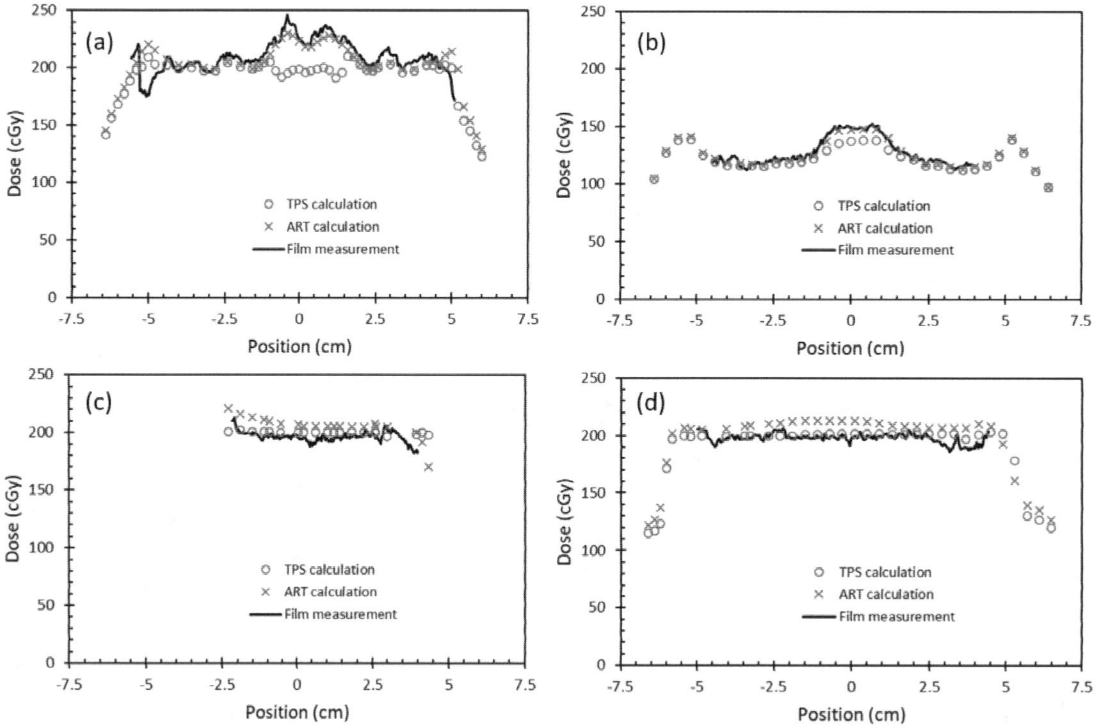

Figure 2. Film results (solid line) compared with dose calculations from TPS (circles) and ART system (crosses): (a) longitudinal profiles on top of Virtual Water, directly beneath the 3D-printed breast, (b) longitudinal profiles within Virtual Water, 5 cm posterior of the 3D-printed breast, (c) transverse profiles (anterior-posterior) through the centre of the 3D-printed breast, and (d) lateral profiles through the centre of the 3D-printed breast.

Figure 2(a) illustrates how the treatment planning system achieved a homogeneous dose profile in the tissue-equivalent material directly beneath the model breast, by increasing the photon fluence through the metallic component of the implant. Figure 2(a) also shows that the TPS apparently over-estimated the beam attenuating effect of the metal CO_2 container (through the central $\pm$ 1.0 cm of the profile), with both the film measurement and the ART calculation showing doses greater than the TPS-calculated dose in this region. Similar results, with the film measurement and the ART calculation both exceeding the TPS-calculated dose in the region where the beam is transmitted through the metal CO_2 container, are apparent even in the approximately tissue-equivalent material 5 cm posterior to the model breast (figure 2 (b)). These results were likely caused by the TPS over-estimating the beam attenuating effect of the stainless steel CO_2 container due to the kV CT images showing the container as larger than its true physical size [13]. This apparent geometric difference may have been as large as 0.2 cm in all directions [13], which sums to 0.8 cm in the anterior-posterior direction, for this hollow gas container.

Figures 2(c) and (d) show profiles that were measured and calculated in a transverse plane, within the 3D-printed model of the breast and TTE. These results indicate that the TPS provides a more reliable indication of the dose throughout the gas inside the implant, with the ART calculation noticeably exceeding the dose calculated by the TPS and measured using film. The disagreement between the ART calculation and the TPS calculation increases in regions closer to the CO_2 container, possibly due to increasing levels of noise (appearing as localized density variations) within the MV CT image that was used by the ART calculation.

4. Conclusions

In tissue-equivalent media downstream of the implant, there was comparatively good agreement between the measured and calculated dose distributions, except in regions where the dose distribution was affected by the stainless steel CO_2 container, where the TPS calculation may have over-estimated the volume occupied by stainless steel. It should be possible to avoid this discrepancy through the use of an extended CT range and a metal artefact reduction algorithm when reconstructing kV CT images of gas-filled TTEs, to minimize imaging artefacts and reduce reliance on density over-rides.

By contrast, film measurements showed that the TomoTherapy TPS provided a more accurate calculation of dose within the air inside the model TTE, with the TomoTherapy ART system over-estimating dose in this region. For each patient with a gas-filled TTE, it is therefore advisable to contour an additional treatment target, which includes only the patient's targeted tissues and excludes all parts of the implant, for use when evaluating and comparing the results of TPS and ART dose calculations.

If the above two precautions are carefully applied, then it should be possible to achieve and act upon accurate dose calculations from the Tomotherapy TPS and the TomoTherapy ART system, when delivering adaptive radiotherapy treatments to patients with gas-filled TTEs.

5. References

[1] Welsh JS, Lock M, Harari PM *et al* 2006 *Technol. Cancer Res. Treat.* **5** 465-479
[2] Schirm M, Yartsev S, Bauman G *et al* 2008 *Technol. Cancer Res. Treat.* **7** 425-432
[3] Sterzing F, Kalz J, Sroka-Perez G *et al* 2009 *Technol. Cancer Res. Treat.* **8** 343-352
[4] Kairn T, Khoei S, Markwell TS *et al* 2010 *Phys. Med. Biol.* **55** N533-N545
[5] Ho AY, Hu ZI, Mehrara BJ *et al* 2017 *Lancet Oncol.* **18** e742-e753
[6] Chopra K, Singh D, Hricz N *et al* 2019 *Plast. Reconstr. Surg. Glob. Open* **7** 1-6
[7] Moni J, Saleeby J, Bannon E *et al* 2015 *Pract. Radiat. Oncol.* **5** e1-e8
[8] Tran T, Ding W, Subramanian B 2014 *Int. J. Cancer Ther. Oncol.* **2** 1-6
[9] Kairn T, Crowe SB 2020 *Phys. Eng. Sci. Med.* **43** 347
[10] Kuo JV, Ramsinghani NS, Evans GR *et al* 2013 *Int. J. .Radiat. Oncol. Biol. Phys.* **87** S219
[11] Kairn T, Hardcastle N, Kenny J *et al* 2011 *Australas. Phys. Eng. Sci. Med* **34** 333-343
[12] Dancewicz O, Sylvander SR, Markwell TS *et al* 2016 *Australas. Phys. Eng. Sci. Med.* **39** 1155-56
[13] Kairn T, Crowe SB, Fogg P *et al* 2013 *Australas. Phys. Eng. Sci. Med.* **36** 209-217

Journal of Physics: Conference Series **2167** (2022) 012013 doi:10.1088/1742-6596/2167/1/012013

Three-dimensional reproducibility assessment of radiochromic ClearView™ gel in custom vessel

Kevin Jordan[1,2,3], Lu Xu[1] and Rob Barnett[1,2,3]
[1]Department of Medical Biophysics, Western University, London, ON, Canada
[2]Department of Oncology, Western University, London, ON, Canada
[3]London Regional Cancer Program, London Health Sciences Centre, London, ON, Canada.

Email: kevin.jordan@lhsc.on.ca

Abstract. Samples of the radiochromic hydrogel ClearView™ were characterized for 3D dosimeter performance in custom vessels and read with a modified Vista16 optical cone beam scanner as part of a small field dosimetry study. Uniform dose responses and small field depth doses revealed each sample was unique. As the gels aged over several weeks at 4 °C, the dose sensitivity decreased. Reconstructions of uniform dose distributions revealed unique features near the vessel interface. The central sub-volumes ranged from uniformity to within 2% throughout the sub-volume to a 12% gradient from base to top of gel. Comparisons of small field depth doses revealed that if the central volume had a uniform dose response, no correction was required. However, if a gradient in sensitivity from bottom to top was measured a correction may or may not be required. It is suspected that the filling protocol of the custom vessels is a factor in the observed variations in 3D dose response.

1. Introduction

In order to perform accurate 3D dosimetry the entire measurement protocol needs to be calibrated. This includes evaluating the dosimeter, vessel, readout scanner and data processing algorithms. During a study of motion effects on 3D small field dosimetry, repeat irradiations were conducted and the spatial uniformity of dose response is reported for set of four samples.

ClearView™, a radiochromic hydrogel from ModusQA, was selected for this study because of its acceptable dose sensitivity at 20 Gy maximum dose, no diffusion and sufficient optical transmission for 15 cm diameter samples [1-3]. The experiments were refined over several irradiations conducted over a year and the later samples are the focus of this report. The experiment included four small beam irradiations, consisting of a static gel as sample reference and 3 repeat irradiations with the gel moving. The cylindrical gel was irradiated through the flat base and each beam was delivered to a separate quadrant. This sequence provided an internal reference, the static beam, for each experiment. However, this approach assumes that each sample has a uniform dose response. In order to verify isotropic dose response, samples were irradiated with a uniform dose before or after the small field motion experiment.

2. Methods

ModusQA prepared the ClearView batches and filled the custom polyethylene terephthalate (PETE) vessels, constructed by one of the authors. These custom vessels have higher optical quality walls than the current blow-molded jars provided by ModusQA. However, the vessels have a much smaller hole for filling and had different heights compared to the blow-molded jars. This resulted in filling procedures that were different from those optimized by ModusQA for the jars. Optical CT scanning was performed with a Vista16 cone beam scanner provided by ModusQA. The scanner was modified, in-house, by placing a 15 mm diameter aperture at the source to minimize stray light [4]. Early scans, at 530 nm, were scaled for direct comparison with the later scans performed at 590 nm, based on the absorption spectrum. One thousand projections were acquired for both the reference and data scans with a 30 minute post irradiation delay for image development. Reconstructions were performed with iterative OSC-TV algorithm and default parameters, implemented by ModusQA in VistaScan program with 0.5x0.5x0.5 mm^3 voxels. Central axis attenuation coefficients were averaged over 1.5x1.5x0.5 mm^3 for figures 1 and 2. Monte Carlo calculations were performed with EGSnrc code and results agreed with measured depth doses for pinpoint ion chamber.

The samples were received within two days of preparation and stored at 5 °C in the dark until radiation. Samples were then warmed in a water bath for 4-6 hours and kept at 20 °C for the experiment. In this study samples are identified by radiation date. *May14* was same sample as *May2* but stored at 20 °C in dark since May 2nd experiment and was two months old. *Sept16* (one week old) and *Sept30* samples were from a later batch. *Sept30* vessel was 15 cm diameter by 8 cm height with 0.5 mm polyethylene terephthalate glycol (PETG) ends and other vessels were 10 cm diameter by 12 cm height with flexible 0.25 mm thick PETE film ends. Vessel cylinder sides were constructed of the same 0.25 mm PETE film. Irradiations were performed with a Varian 2100 IX series linear accelerator; 6 MV, 600 MU/min. Small field irradiation details: gel base facing beam at 90 cm source to surface distance, jaw sizes 1.4x1.4 cm, 2642 MU per beam (~20 Gy at 50 mm depth). Uniform dose irradiation details: 10 Gy, sample in 20x20x20 cm water tank, gel center at 100 cm isocenter, 4-field delivery.

3. Results

Reconstructions of the uniform dose irradiations revealed the samples differed from each other in their dose response. Sub-volumes along the ends and cylinder walls showed local variations in reconstructed attenuation coefficients. For this reason, the outer 1 cm of gel is excluded from this study. The central volume showed radially uniform response to <5%. However, *May14* sensitivity linearly increased ~12% from base to top, *Sept16* linearly decreased ~12% from base to top, *Sept30* was uniform <2% throughout volume and *May2* did not receive a uniform dose. Note, *Sept30* uniform dose experiment showed similar response in central volume but different spatial responses near perimeter compared to the same vessel filled with a previous batch. Indicating the vessel filling protocol has measurable impact, especially in sample perimeter. Reconstructions also reveal high contrast artifacts that are due to 'short strings of solid gel' dispersed throughout the samples.

Figure 1, is a plot of reconstructed attenuation coefficients for the static beam central axis from each experiment. *Sept16* was the most sensitive and a decrease in sensitivity due to storage time was seen with *Sept30* and *May2* curves. *May14* scan reveals dose sensitivity increased from *May2* due to interim storage at 20 °C. From other data sets, sensitivity batch to batch is similar when comparing samples that have been in 4 °C storage for longer than a month. The hollow marker curves show attenuation coefficients corrected linearly from the base to the top for changes in sensitivity recorded for the *Sept16* and *May14* samples. No spatial correction was required for the *Sept30* experiment and the *May2* scan was plotted using the *May14* correction for comparison. Note, the *Sept30* vessel was shorter, providing less depth dose data. In figure 2, the curves are normalized to a depth of 50 mm, to minimize impact of spatial variations in sensitivity near the sample's outer sides. The curves which have better agreement with Monte Carlo depth doses are uncorrected *Sept30*, *Sept16* and corrected *May14*.

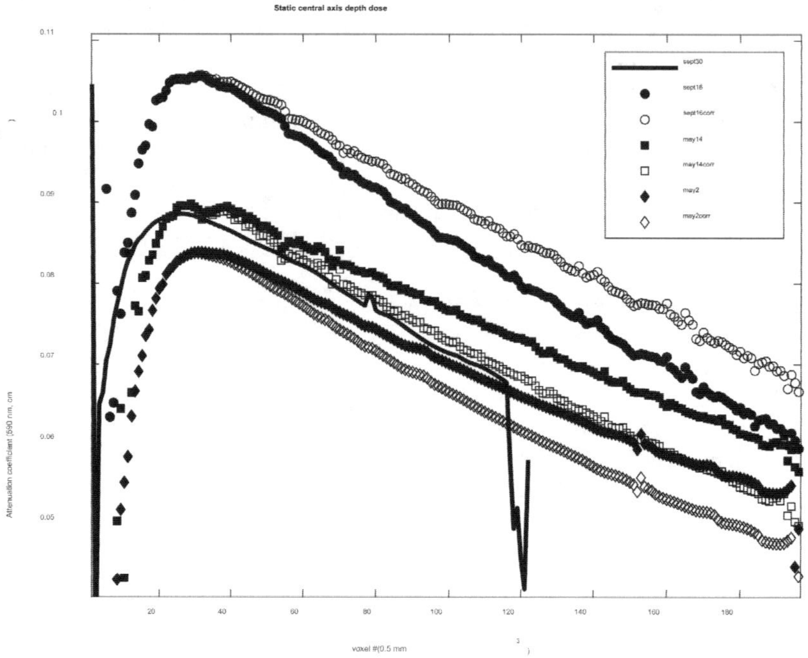

Figure 1. Plot of central axis attenuation coefficients versus slice #, solid line or markers. Correcting bottom to top change in sensitivity, open markers. Note, vessel height *Sept30* 8 cm and others 12 cm.

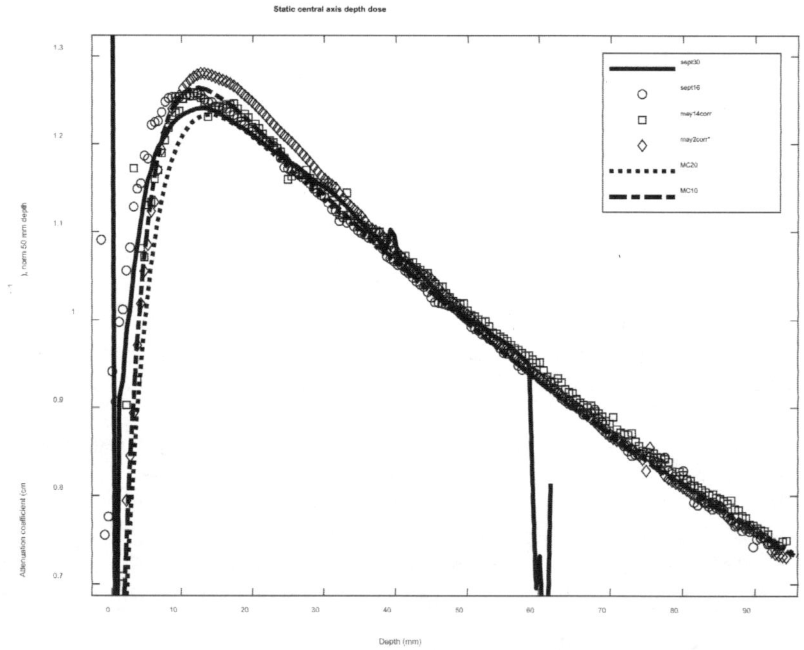

Figure 2. Plot of central axis attenuation coefficients vs Monte Carlo calculated depth doses. Curves normalized at a depth of 50 mm. Note *May2corr* curve generated with correction from *May14* irradiation as a surrogate.

May2 corrected with *May14* data is also presented. The *Sept30* results demonstrate that 3D dosimetry is possible, but the *Sept16* spatial response indicates that a correction should be applied. However, the comparison with Monte Carlo depth doses indicates no correction is required. The *May14* result shows the opposite effect, namely a spatial correction is required to get agreement. While *May2* corrected indicates *May14* correction is applicable from top to 3 cm above base.

4. Discussion

ClearView™ gel has the potential for accurate 3D dosimetry as shown by agreement between Monte Carlo calculations and measured depth doses. However, early results with custom vessels show that each sample is unique and a correction for spatial variations in dose sensitivity are sometimes required. Further experimentation is required to identify the causes for these spatial variations and why corrections may or may not be required. Similar effects have not been reported with more aged samples in the Vista jars but decreases in sensitivity and spatial variations were observed previously during the first month [5]. It is anticipated that an optimized filling procedure for these custom vessels is needed to obtain reproducible 3D responses. Another potential factor is the custom vessels. The *Sept30* vessel was constructed with 0.5 mm thick PETG ends vs the 0.25 mm PETE ends on the other vessels. The thicker ends provided stiffer vessels and limited flexing when filled with ClearView™. It had been observed that vessels with air head spaces could develop a layer of liquid on their surface and along the sides if tipped on side or upside down for irradiations through the base end. A liquid between the vessel wall and gel may have an impact of transmission images. ClearView™ gel does not adhere to PETE in contrast to gelatin hydrogels which bond to the plastic interface. It also appears to shrink slightly with aging and this could introduce an optical effect at the vessel wall. These differences may be a factor in reproducibility for the results presented here. Once this base to top gradient in dose response is minimized, smaller effects such as impact of aging and oxygen gradients can be identified and addressed.

5. Acknowledgments and Disclosures

A Collaborative Health Research Project (CHRP no. 356860) grant from the Natural Sciences and Engineering Research Council of Canada (NSERC) and the Canadian Institutes of Health Research (CIHR), with Modus QA as an industry partner provided the Vista16 optical CT scanner and the gel samples. KJJ is signatory to a licensing agreement, concerning optical CT, between Western University, Lawson Health Research Institute, and Modus QA.

6. References

[1] Penev K I, Wang M, Mequanint K 2017 *J. Phys. Conf. Ser.* **847** 012048
[2] Ascencion Y, Dietrich J, Mequanint K *et al* I 2017 *J. Phys. Conf. Ser.* **847** 012049
[3] Colnot J, Huet C, Gschwind R *et al* 2018 *Phys. Medica* **52** 154-164
[4] Dekker K H, Hazarika R, Jordan K J 2018 *Phys. Med. Biol.* **63** 21NT02
[5] Hazarika R *et al* 2018 *Med. Phys.* **45** SU-I-GPD-T-321

Journal of Physics: Conference Series **2167** (2022) 012014 doi:10.1088/1742-6596/2167/1/012014

Effects of PVA-GTA-I radiochromic gel dosimeter components on optical dose-response

Shin-ichiro Hayashi[1], Kaoru Ono[2], Keisuke Fujino[2], Ryosuke Kurihara[2]
[1]Department of Clinical Radiology, Faculty of Health Sciences, Hiroshima International University, Higashi-Hiroshima, Hiroshima 739-2695, Japan
[2]Hiroshima Heiwa Clinic, Hiroshima, 730-0856, Japan

Email: rin@hirokoku-u.ac.jp

Abstract. In recent years, a novel radiochromic gel dosimeter was developed that utilizes the color development of a polyvinyl alcohol-iodide (PVA-I) complex. In this study, we explore the effects of different iodide salts (LiI, NaI, KI, CsI, NH_4I, CaI_2, and ZnI_2) and PVAs with different degrees of polymerization (500, 1000, and 1500) and saponification (80, 88, and 98 mol%) were investigated on a PVA-GTA-I gel dosimeter using PVA that was chemically crosslinked with glutaraldehyde (GTA) as a matrix. The results showed that these substitutions had negligible effect on dose-responses, such as sensitivity and dose-rate independence.

1. Introduction

Recently, we developed a novel radiochromic PVA-I gel dosimeter [1, 2]. The dosimeter utilizes the red color development due to the complexation of polyvinyl alcohol (PVA) with iodide (I_3^-) formed by irradiation in an aqueous solution. The PVA-I gel is composed of partially saponified PVA, potassium iodide (KI), fructose as a reductant, gellan gum as a gelling agent, and distilled water. This gel dosimeter has excellent dose characteristics such as high sensitivity, dose rate independence, and a wide dose range. In particular, the PVA-I gel dosimeter can be decolored (initialized) by heating, owing to the reducing nature of fructose, resulting in its reusability. In this gel dosimeter, PVA acts as a host for iodide ions to form a complex. Therefore, a gelling agent is required for the gelation of the solution. The gellan gum used in the PVA-I gel dosimeter is solidified by physical cross-linking with cations.

On the other hand, a Fricke-type gel dosimeter using PVA chemically crosslinked with glutaraldehyde (GTA) as a matrix was reported and named PVA-GTA Fricke gel dosimeter [3, 4]. Employing the PVA-GTA matrix in the Fricke gel dosimeter resulted in high transparency, low diffusivity of ferric ions created by irradiation, and higher dose sensitivity. If the PVA-GTA matrix can be applied to the PVA-I gel dosimeter, then no gelling agent would be required, resulting in higher transparency. Therefore, a PVA-I gel dosimeter using a PVA-GTA matrix (PVA-GTA-I gel dosimeter) was developed [5, 6].

In the present study, we investigated the effects of different iodides (LiI, NaI, KI, CsI, NH_4I, CaI_2, and ZnI_2), the average degree of polymerization (DP: 500, 1000, and 1500), and the degree of saponification (DS: 80, 88, and 98 mol%) of PVA on the dose response of the PVA-GTA-I gel dosimeter.

Table 1 Components of the PVA-GTA-I gel dosimeter prepared in this study. The standard composition is marked in bold. M is molarity (mol/1000gH_2O).

Base solution									
PVA	Average degree of polymerization			**500**	1000	1500			10 wt%
	Degree of saponification (mol%)		80	**88**	98				
Water									90 wt%
Additives									
Iodide		LiI	NaI	**KI**	CsI	NH_4I			100 mM
							CaI_2	ZnI_2	50 mM
Reducing sugar				**fructose**					100 mM
Cross-linker				**GTA**					10 mM
Proton generator				**GDL**					100 mM

2. Materials and methods

2.1 Gel preparation

The composition of the PVA-GTA-I gel dosimeters manufactured in this study is listed in Table 1. Iodide salt, fructose, GTA, and glucono-δ-lactone (GDL) were added to a 10 wt% PVA solution prepared in advance. GDL was employed as a proton generator to catalyze crosslinking. All the reagents were completely dissolved using a magnetic stirrer at room temperature (15-20 °C). The obtained mixed solution was divided into PMMA cuvettes (4.5 mL, 1 cm path length). Before irradiation, the samples were initialized and solidified by annealing at 45 °C for 12 h in an incubator.

2.2 Irradiation

Similar to a previous study [2], all samples were irradiated using a 6 MV X-ray from a medical linear accelerator (Varian/BrainLAB Novalis Tx, USA) at a depth of 5 cm in solid water phantom at room temperature (20-25 °C). Doses of up to 10 Gy were delivered to the samples, with an average dose rate fixed at 600 cGy/min. Different dose rates of 100, 200, and 400 cGy/min were used to investigate the dose rate dependence.

2.3 Measurement

The absorption spectra of the samples were also measured as described in a previous study [2], one day after irradiation, using a UV-Vis spectrophotometer (UV-1600PC, Shimadzu, Japan). Absorbance (Abs) was calibrated using distilled water. The change in absorbance (ΔAbs) was calculated by subtracting the absorbance of the non-irradiated sample from that of the irradiated sample.

3. Results and Discussion

3.1 Absorption spectra

Figure 1(a) shows the PVA-GTA-I gel dosimeters irradiated with up to 10 Gy. As the absorbed dose increased, their appearance gradually changed from colorless to red, similar to that observed for the PVA-I gel dosimeters [1, 2]. Fig. 1(b) shows the absorption spectra of PVA-GTA-I gel dosimeters. Single peaks centered at approximately 486 nm were exclusively observed. Therefore, the absorbance at 486 nm was used in subsequent investigations to determine the dose responses. However, as shown in Fig.1(c), the spectra of the PVA-GTA-I gel containing PVA with a DS of 98 mol% were different from the others. It displays a remarkable absorbance at 358 nm due to the free iodide ion (I_3^-) that did not form a complex with PVA, while only a small absorption increase was observed at 486 nm. In addition, Fig.1(d) shows the background absorption spectra of PVA-GTA-I and PVA-I gel dosimeters before irradiation (immediately after initialization). The background absorbance (Abs = 0.002) of PVA-GTA-I at 486 nm was substantially lower than that of PVA-I (Abs = 0.040), resulting in higher transparency (lower background).

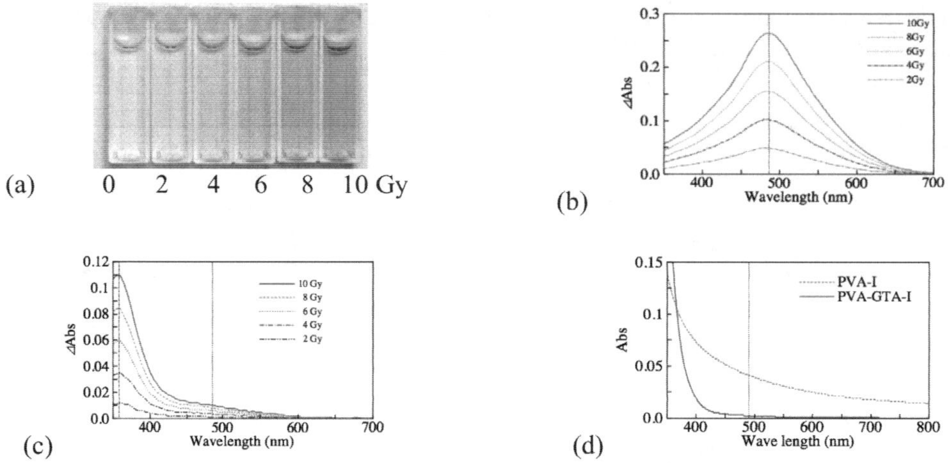

Figure 1 (a) PVA-GTA-(Na)I gel dosimeters irradiated up to 10 Gy. (b) Absorption spectra of PVA-GTA-(Na)I gel dosimeters irradiated up to 10 Gy. (c) Absorption spectra of PVA-GTA-I gel dosimeters containing PVA with 98 mol% DS irradiated up to 10 Gy. (d) Absorption spectra of non-irradiated PVA-GTA- I gel and PVA-I gel dosimeters immediately after initialization (annealing).

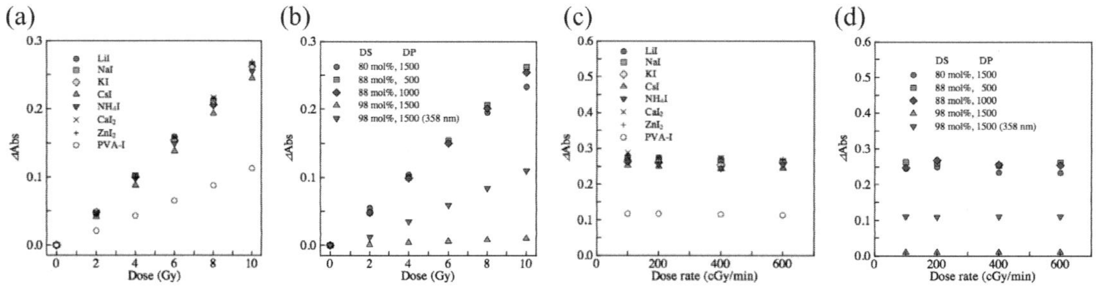

Figure 2 (a) and (b) show the dose responses of the PVA-GTA-I gel dosimeters containing different iodide salts and PVAs with different DSs and DPs, respectively. (c) and (d) represent the dose-rate dependence of the absorbance at the same dose (10 Gy) in the corresponding gel dosimeters. White circles (○) show the response of the PVA-I gel dosimeter for comparison [2]. The data points, except for PVA-I and 98 mol% DS, nearly superimpose each other.

3.2 Dose-responses

3.2.1 Effect of Iodide type

Figure 2 (a) shows the dose response of the PVA-GTA-I gel dosimeters containing different iodide salts. The sensitivities were similar, regardless of the type of iodide salt used. The results indicated that the cation does not affect the dose response of a high-energy photon beam used in radiotherapy, although heavy atoms, such as Cs, could affect the sensitivity in the lower energy regions due to photoelectric absorption. In addition, the sensitivities of the PVA-GTA-I gel dosimeters were approximately 2.5 times higher than those of the PVA-I gel dosimeter [2].

3.2.2 Effect of degree of polymerization (DP) and saponification (DS) of PVA
Figure 2 (b) shows the dose responses of PVA-GTA-I gel dosimeters fabricated with PVAs of different DPs and DSs. The results demonstrated that the sensitivity was only slightly affected by DP and DS, except for that containing PVA with a DS of 98 mol%. This indicates that the number of acetyl residues on PVA is not sufficient to form the maldistributed groups required to form a complex with triiodide ions. The dose response at 358 nm, corresponding to free iodide, showed adequate linearity and sensitivity. This is consistent with a previous study on dosimetry using an aqueous KI solution [7]. However, dosimetry using a PVA-I complex provides higher sensitivity and visibility than dosimetry using a KI aqueous solution. On the other hand, a 10 wt% PVA base solution containing PVA with DS of 80 mol% or a higher DP exhibited higher viscosities than the other PVA aqueous solutions. Therefore, the authors recommend fabricating a gel dosimeter using PVA with lower viscosity, DP of 500–1000, and DS of 88 mol% for the uniformity of the gel.

3.3 Effect of dose rate dependence
Figures 2 (c) and (d) show the absorbance of PVA-GTA-I gel dosimeters containing various components irradiated at 10 Gy at different dose rates. It was shown that dose-rate-independence, which is one of the favorable characteristics of PVA-iodide-based gel dosimeters [1, 2], is also preserved by the substitution of different iodide salts or PVAs in the PVA-GTA-I gel dosimeter.

4. Conclusion
In the present study, the effects of various components on the dose response of the PVA-GTA-I radiochromic gel dosimeter were investigated. The difference in the iodide salts or PVAs had negligible effect on the dose properties of the PVA-GTA-I gel dosimeter, except for the gel dosimeter containing 98 mol% DS of PVA. These results could help optimize the selection of the ingredients for gel dosimeters. Moreover, the PVA-GTA-I gel dosimeter demonstrated higher transparency and sensitivity than those of the PVA-I gel dosimeter. Based on these results, the PVA-GTA-I gel dosimeter exhibits the potential to be a more effective dosimetry tool for radiotherapy. However, further optimization of the composition and evaluation of its fundamental properties, such as temporal, spatial, and thermal stabilities, are still required. Efforts are currently underway to address these issues.

5. Acknowledgment
This work is supported by JSPS KAKENHI (Grant Numbers 17K09072 and 20K08039).

6. References
[1] Hayashi S *et al* 2019 *J. Phys.: Conf. Ser.* **1305** 012031
[2] Hayashi S *et al* 2020 *Radiat. Meas.* **131** 106226
[3] Gallo S *et al* 2019 *Radiat. Phys. Chem.* **160** 35-40
[4] Marini A *et al* 2017 *Radiat. Meas.* **106** 618-621
[5] Taño J E *et al* 2019 *J. Phys.: Conf. Ser.* **1305** 012034
[6] Taño J E *et al* 2020 *Radiat. Meas.* **134** 106311
[7] Rahn R O *et al* 2002 *Appl. Radiat. Isot.* **56** 525-534

Marker trajectory assessment in optical cone beam computed tomography scanner geometry

Gareth Marks[1] Jaryd Christie[2] and Kevin Jordan[1,2,3]
[1]London Regional Cancer Program, London Health Sciences Centre, London, ON, Canada
[2]Department of Medical Biophysics, Western University, London, ON, Canada
[3]Department of Oncology, Western University, London, ON, Canada

Email: kevin.jordan@lhsc.on.ca

Abstract. Measured trajectories of markers on vessel wall and beads on a plumb are compared with calculated trajectories for transmission images obtained with Vista16™ optical cone beam scanner. Effective pixel size was 0.11 mm. An algorithm was developed to measure trajectories for the plumb line phantom and a semi-automatic version was required for the gel filled vessels due to overlap of markers and seam projections at maximum lateral positions. It was found that for 15 cm diameter vessels, the vessel edge was not visible due to divergence in cone beam geometry. Accounting for this effect was necessary in order to remove artefacts from calculated vessel trajectory. Displacement differences were found to be less than 0.1 mm for the plumb line, 0.3 mm for a beads and 1 mm for markers on gel filled vessel. Minimization of the vessel marker trajectory deviation will provide a more accurate approach to refractive index optimization. No optical aberrations were detected from the trajectory analysis and manufacturer scanner alignment and geometrical calibrations were independently verified.

1. Introduction

The Vista16™ optical cone beam computed tomography (CT) scanner from ModusQA has a 16 cm field of view. Adding a 1.5 cm diameter aperture at the source reduces stray light in transmission images [1]. One consequence of reducing stray light is the need for higher optical quality cylindrical plastic vessels for 3D radiochromic hydrogel dosimetry. While initial images and 3D reconstructions of coloured solutions with in-house constructed custom vessels are promising, inspection of pre and post scans often reveal displacements due to limitations of the custom mounts for these custom vessels. Improvements in custom vessel design and reproducible mounting is ongoing. However, image registration is a powerful method to improve CT reconstruction without placing prohibitive demands on the imaging system. An example of this is the IsoCal phantom and imaging protocol used to characterize the kilovoltage and megavoltage imaging systems on Varian medical linear accelerators [2]. X-ray cone beam CT scanner geometry calibration has been addressed by many researchers [3-5]. Often the calibrations involve analysis of small spheres or wires at optimum locations in a custom calibration cylinder. Preliminary results of writing millimeter wide lines into a radiochromic and photochromic genipin hydrogel have been presented at a previous IC3DDose conference [6].

In this study 1 mm black ink marks were placed on walls of custom gel cylinders and 2 mm beads were attached to a plumb line. Trajectories were measured from optical cone beam CT scans of these phantoms. The plumb line follows the same trajectory as the ink mark on the vessel wall. The trajectories in the horizontal plane containing the optical axis are also in the sinogram for this plane. Comparisons of measured and calculated trajectories provide information on the system's geometry and deviations due to aberrations. For example, trajectory deviations should determine: rotational symmetry of rotary bearing for sample mount, accuracy of rotational increments, net spherical aberration, refractive index uniformity of gels and impact of differences in refractive indices of gel, vessel wall and outside liquid.

2. Methods

Black ink marks were added to a custom polyethylene terephthalate (PETE) vessel, 50 and 100 mm above the bottom and adjacent to the single, 1 mm wide, black, wall seam, see figure 1. Vessel wall thickness was 0.25 mm and the outer diameter was 148.2 mm +/-0.2 mm. Optical CT scanning was performed with a Vista16 cone beam scanner provided by ModusQA modified to a 1.5 cm diameter source. A 12% propylene glycol aqueous solution provided optimum field of view for ClearView™ gel in this vessel. The pixel size corresponded to 0.11 mm in the rotation axis plane. The vessels rotate counter clockwise when viewed from above and are mounted with the seam at the source side on the optic axis. One thousand projections were acquired over 360 degrees per scan. Images were exported from VistaScan as PNG files and processed in MATLAB. A plumb with small beads was attached to outer diameter of an end disk used for custom vessel fabrication. The plumb line, (0.3 mm diameter, black fishing line), allows measurements of reference trajectories in solution without the aberrations due to the vessel and gel. The beads were constructed from short lengths of black plastic tubing crimped to the line.

The optical axis was determined by recording images of a transparent ruler placed vertically in the filled aquarium, resting on the bottom at the camera side and source side. Similarly, the ruler was placed horizontally, resting on the side of the aquarium again at the camera side and source side and recording images. Inspection of the images determined the row and column that did not diverge. The intersection was the optical axis. This value was within 2 pixels of the center pixel of the sensor. An image of the aquarium half-filled provided another measure of the height of the optic axis and horizontal alignment and up down tilt of the camera. No adjustments of the camera position from manufacturer were performed. Aquarium dimensions were measured with a machinist's vernier, <0.5 mm uncertainty. Rotation axis was measured by several methods and was found to be within 2 pixels, 0.2 mm of the optic axis.

The most general form of the algorithm used to detect marker trajectories uses a semi-manual approach, where the user periodically inputs the new region of interest. For data sets with lower noise, a fully automated approach was used. On a given frame, the algorithm searches a particular region of interest for a connected group of pixels with brightness values below a particular threshold. The region of interest image is binarized about this threshold. Depending on details of the marker being searched for, other operations are then performed. For instance, for a bead hung on a plumb line, the image is eroded by a suitable structural element to remove the line. If part of the opaque vessel seam or other artefact is present in the region of interest, the program can account for this when selecting the proper region. The geometric center of the final pixel group is then calculated, and the new region of interest centered around that point in the next frame. For markers placed near an opaque seam, the algorithm switches to tracking the seam at points where the marker approaches the edge of the image. Since the seam is vertical in the image, this allows for column number data to be extracted for more frames, though obviously row number data cannot be gained this way. The row number is thus linearly interpolated over in these regions.

Once the trajectories have been detected in projected space, they are transformed into physical space for comparison with theoretical values. The vessel diameter, aquarium thickness (parallel to the optic axis) and location of the rotation and optic axes are measured. The divergence of the system, which could not be reliably ascertained from physical measurements, was selected by forcing frames where a rotating plumb line are calculated to appear furthest left and right with the frames where this is actually observed. With these parameters, the actual trajectory of a marker in physical space is calculated and compared to theoretical values assuming perfectly-cylindrical vessels and no unforeseen optical effects.

Obviously, this displacement is highly sensitive to the value used for the vessel diameter. It is also extremely sensitive to the divergence. The limit of 1000 total frames in a rotation, because of the ultimate method used to calculate divergence, limited the precision with which the divergence could be measured, and thus furnished perhaps the largest source of uncertainty. The displacement was, surprisingly, much less sensitive to slight alterations in the values used for aquarium thickness and rotation axis position, though knowledge of the pixel where the optic axis lay in the image was crucial for removing distortion.

3. Results

Figure 1, shows transmission image of vessel containing an irradiated gel with marker positions identified by arrows. This projection corresponds to the markers and seam on the camera side of the vessel in the vertical plane of the rotation axis. Image window and level were adjusted to make makers more visible. The side and bottom edges correspond to the camera side of the aquarium and the corner tabs are from the source side. These edges are used for autocalibration of the magnification. No significant optical distortion inherent to the scanner geometry was found. The plum line with beads and weight suspended in solution is seen in right panel. The plum line is at the same radius as the vessel wall of the left panel. Figure 2, is a plot of the difference between measured and calculated position of marker or bead versus projection number. The first projection corresponds to the marker or bead in the vertical plane of the optic axis on the camera side of the trajectory. The trajectory differences for each bead on the plumb line had a unique pattern and are thought to be due to asymmetries in the bead shape, kinks in line from crimping beads and swaying in the line during scan rotation. This is consistent with line trajectories near the beads having maximum differences 50% smaller (1 pixel difference) than the beads (3 pixels). However, larger differences associated with gel filled vessel were found. The distortion approached 1 mm (approximately ten pixels in images) near the edge of the projection. Note the differences are less for the source side of the trajectory. This is expected since the calculation assumes single uniform refractive index ignoring vessel and gel. It appears differences are primarily due to a slight mismatch between the refractive index in the gel and that of the surrounding propylene glycol solution in the scanner, rather than effects due to the interface with the vessel. In particular, where the refractive index is too low outside the vessel, parts of the vessel near the projection's edge appear further from the optic axis, and where it is too high they appear closer to the optic axis. Also, comparison to datasets using a custom vessel 10 cm in diameter further suggest that the distortion depends more on the distance from the optic axis than on the angle at which incident light rays meet the vessel.

Figure 1. Projection 1 images: (left) markers along custom vessel seam indicated by arrows and filled with irradiated ClearView™ gel, seam at source side, (right) beads on plumb line, at source side. Vista16 aquarium with 12% propylene glycol water solution.

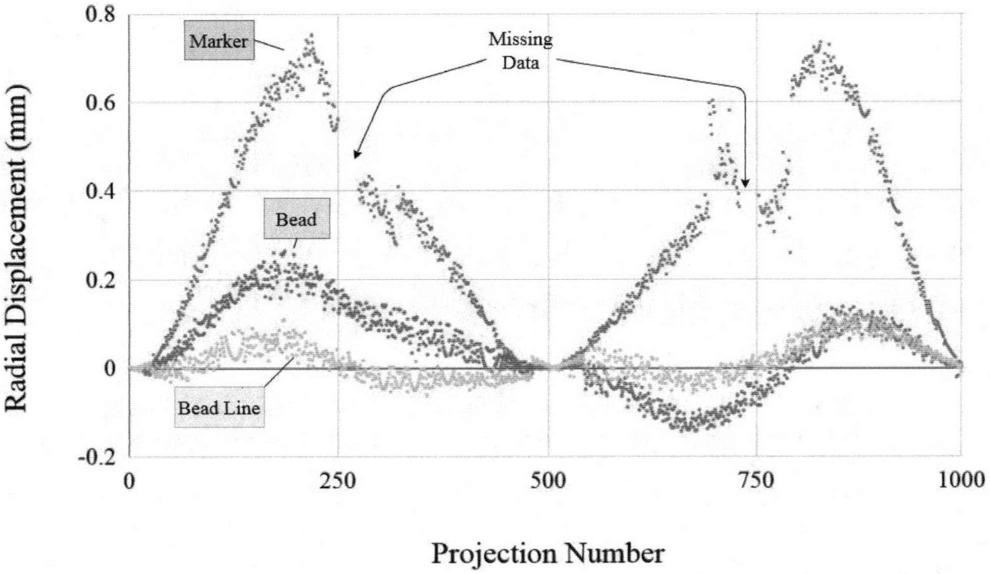

Figure 2. Plot of trajectory position difference between projection image and calculation. Projection range 250 to 750 corresponds to camera side of trajectory. Missing data ranges correspond to marker overlapping vessel seam at edge of vessel. Marker is upper marker of figure 1, gel vessel. Bead is second from top on plumb line. Bead line is a segment beside bead.

4. Conclusion

Small marks on the vessel walls can be used to measure trajectories at different heights relative to the optic axis. For custom vessels with a dark seam and adjacent markers overlapping positions occur near maximum lateral distance in projection plane. Both the dark seam and small marks on vessel wall can be used to measure trajectories and provide geometrical data for accurate optical CT reconstructions. For example, differences between measured and calculated trajectories can be used to optimize the refractive index of the solutions. Trajectories differences can also provide a measure of aberrations in transmission images. It is anticipated that trajectories may be helpful for assessing optical performance of different gels and vessels.

5. Acknowledgments and Disclosures

A Collaborative Health Research Project (CHRP no. 356860) grant from the Natural Sciences and Engineering Research Council of Canada (NSERC) and the Canadian Institutes of Health Research (CIHR), with Modus QA as an industry partner provided the Vista16 optical CT scanner. KJJ is signatory to a licensing agreement, concerning optical CT, between Western University, Lawson Health Research Institute, and ModusQA.

6. References

[1] Dekker K H, Hazarika R, Jordan K J 2018 *Phys. Med. Biol.* **63** 21NT02
[2] Gao S, Du W, Balter P *et al* 2014 *JACMP* **15** 164-181
[3] Cho Y, Moseley D J, Siewerdsen J H *et al* 2005 *Med Phys* **32** 968-983
[4] Ferrucci M, Leach R K, Giusca C *et al* 2015 *Meas. Sci. Technol.* **26** 092003
[5] Jacobson M W, Ketcha M D, Capostagno S *et al* 2018 *Phys. Med. Biol.* **63** 025030
[6] Jordan K 2013 *J. Phys.: Conf. Ser.* **444** 012039

Investigation of target dose conformity using normoxic polymer gel dosimetry techniques: A clinical example of 12th thoracic vertebrae SBRT treatment with VMAT

A Venning[1], M Mundayadan Chandroth[1], C Morgan[2] and M Roberts[2]
[1]Mid-North Coast Cancer Institute, Port Macquarie Base Hospital, Port Macquarie, NSW, Australia
[2]Mid-North Coast Diagnostic Imaging, Highfields Circuit, Port Macquarie, NSW, Australia

Email: anthony.venning@health.nsw.gov.au

Abstract. This paper describes the 3D dosimetry verification of a 24Gy in 3 fractions VMAT SBRT treatment delivery to the 12th thoracic vertebrae (T12). With this fractionation regime, the critical isodose value is 17.5Gy to the outer edge of the spinal cord and must not be exceeded. This is a high-risk treatment that could result in severe injury to the patient unless administered precisely. With the use of 3D normoxic polymer gel dosimetry techniques, using MRI readout, a clinical example is presented which shows that the Mid North Coast Cancer Institute's paraspinal program is capable of delivering complex distributions with high doses to within very tight tolerances. This also demonstrates the strength of normoxic polymer gel as a 3D dosimeter.

1. Introduction

With advances in radiotherapy delivery techniques and image guidance, the Mid North Coast Cancer Institute (MNCCI) has recently developed a paraspinal SBRT program. A common treatment delivery regime is 24Gy in 3 fractions to spinal metastases. With this fractionation, a hard dose constraint is placed on the spinal cord which must not be violated, and the PTV, GTV and CTV coverage is compromised, if necessary, to achieve this. Modern commercially available dosimeters do not offer true 3D delivery verification capabilities [1,2]. Patient quality assurance tools, such as the ArcCheck™ which are routinely used in the clinical environment do not measure dose to high precision. Furthermore, for SBRT patient QA, most of the dose is delivered to the centre of the ArcCheck where there is only located a 0D dosimeter such as an ionisation chamber. With QA devices, such as the ArcCheck, all the dose information is captured at the periphery and the complexities of the volumetric dose distributions around the isocentre are ignored. Using 3D normoxic polymer gel dosimetry techniques with MRI readout [3-10], this study presents a 3D dosimetric analysis of a 12th thoracic vertebrae (T12) VMAT SBRT treatment. While the gel does not have the same radiological tissue equivalence of bone, it does have the same radiological tissue equivalence of soft tissue, providing accurate dose calculation and dose information capture [11]. Figure 1 below, shows the treatment case which is presented in this study.

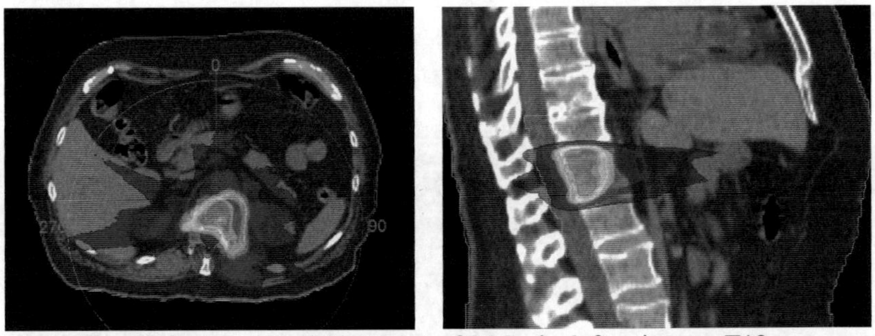

Figure 1. Paraspinal treatment of 24Gy in 3 fractions to T12.

2. Methods and Materials

2.1. Phantom

The gel vial simulating the spine was irradiated in a phantom designed for another concurrent study assessing lung tumour doses [12,13]. This phantom was used due to its human tissue equivalent electron densities as it predominately consists of water. The phantom was made of foam wrapped in 1cm thick Superflab Bolus (NL-Tec Pty Ltd) to simulate chest wall, and the foam was soaked in water to simulate lung tissue electron densities. The correct retention of water was established to be equivalent to that of average human lung tissue electron density. A 47 mm diameter hole drilled out from its central core was used to hold the gel dosimeter firmly in position. The gel container was oxygen proof polyethylene terephthalate (PET) vials of 47 mm dia. x 150 mm hgt.

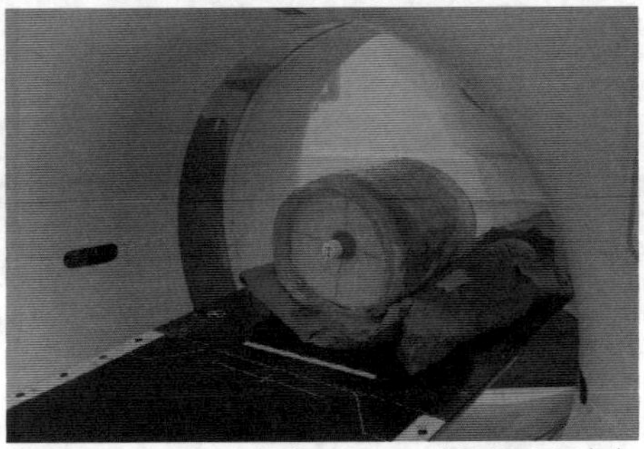

Figure 2. CT of phantom with PET vial.

2.2. Polymer gel method

Normoxic PAG dosimeters were manufactured using techniques described elsewhere [4-6,8,14], with a composition of 5% gelatine, 3% acrylamide, 3% N, N′-methylenebisacrylamide, 2mM hydroquinone, 5 mM tetrakis phosphonium chloride and 89% de-ionised water. A total of 4 PET vials were filled with this gel, 3 used for R_2-dose calibration and 1 used for the T12 treatment plan measurement.

2.3 Treatment planning

CT images of the wet foam with a water filled PET vial were exported to the Monaco TPS. In the TPS, the patient's dataset and phantom were fused together such that the spinal cord and PTV contours could be transferred onto the PET vial and the plan isocentre could be established. The patient's bi-arc VMAT treatment plan was copied onto the CT data set and recalculated at 1mm voxel with 1% STD, retaining MLC beam shapes. As per protocol, for paraspinal SBRT treatments delivered in 24Gy/3# must meet the hard dose constraint on the spinal cord of 17.5Gy and must not be exceeded [15]. In these cases, the PTV, CTV and GTV are to be compromised to achieve this. However, for this dosimetry study, the plan was scaled down as 24Gy in 3 fractions leads to a single fraction dose that would saturate the gel dosimeter. The scaled isodose line of interest in this study that was shaped around the spinal cord was 5Gy as shown in figure 3.

2.4 Irradiation
The phantom was located at the isocentre of the linac with a PET vial filled with water and a cone beam CT was performed and then position matched to the reference data using HexaPOD™. The PET vial with water was then carefully replaced with a vial filled with polymer gel and irradiated as per the scaled VMAT SBRT treatment plan. Using 3 vials the R_2-dose calibration curve was generated using the PDD method and a 6MV beam under TRS-398 reference conditions in water as shown in figure 4.

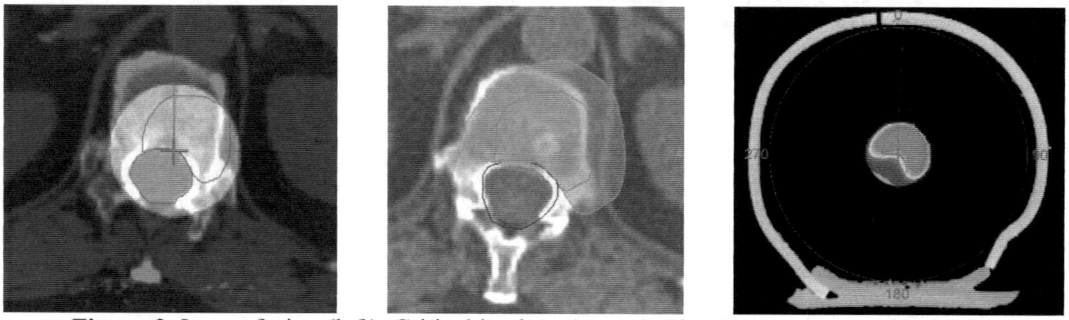

Figure 3. Image fusion (left). Critical isodose (centre). Phantom treatment plan (right).

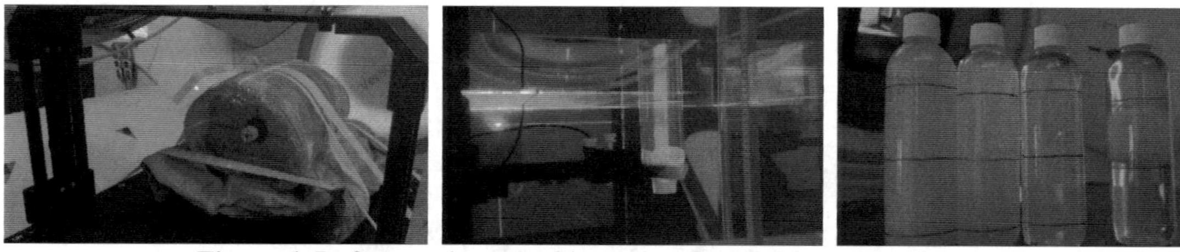

Figure 4. Reference data matching using HexaPOD and irradiation.

2.5 Polymer gel readout
The polymer gels were imaged 5 days post-irradiation using a Siemens Magnetom Vida 3T clinical MRI scanner with head coil. A multi spin-echo pulse sequence with 16 echoes, echo spacing of TE = 40 ms, pixel size of 1 x 1 mm^2 and with 128 phase encoding steps was used to generate T_2 maps [3]. The calibration vials were imaged with single transverse slice at different known dose deposition locations with a slice thickness of 5mm. The gel irradiated according to the T12 treatment plan was imaged in the transverse orientation with 20 adjacent 2mm slice thicknesses (4mm centres) using an interleaved sequence that covered the irradiated volume of the gel dosimeter. As a note, this slice thickness leads to a reduced signal to noise ratio and subsequently produces noisy images. This is due to a limiting time constraint. Ideally, image averaging would be performed but this comes at a significant time efficiency loss.

2.6 Analysis
The MRI DICOM images were processed using MATLAB™ [16]. R_2 maps were generated for each of the 20 slices and converted to dose maps using the R_2-dose calibration curve. A 3D dose matrix was created from the individual dose maps and the 5Gy isodose line was plotted in 3D in MATLAB™ using the isosurface function. The patient's CT dataset was imported into MATLAB™ and the spinal cord contour surrounded by the T12 vertebra was extracted from the dataset. From the fused images of the patient's dataset and the gel phantom, the position of the PET vial that encompassed both the PTV and spinal cord contours was established. The key assessment criteria for this study was to observe if the

5Gy isodose line was shaped around the spinal cord and hence the treatment delivery did not violate the hard 5Gy isodose line constraint in this case.

3. Results and Discussion

3.1. R_2-dose response

Figure 5 shows the R_2-dose response curve with an R_2-dose sensitivity of 0.19 $s^{-1}Gy^{-1}$ in the linear region.

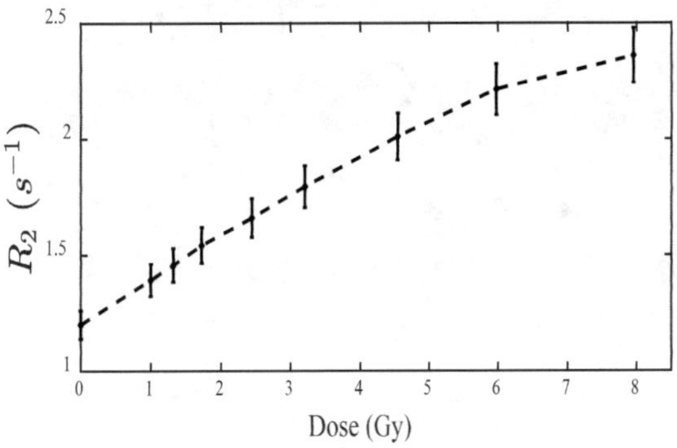

Figure 5

3.2. Gel isosurface and spinal cord

Figure 6 shows the 5Gy (red) isodose sphere of interest wrapped around the spinal cord (blue). At no point does the 5Gy isodose line enter or impinge upon the spinal cord along its length of the treatment region. The 5Gy isodose line is adjacent to the spinal cord to within 1 mm at its closest point which is reassuring inconsideration of how tight the radiation oncologist is pushing the major isodose line of consideration as in figure 3 above. This shows, that the full treatment chain is capable of safely delivering complex dose distributions, and means there can be a greater level of confidence in dose escalation and planning high dose fractionation regimes close to critical organs at risk. There is no other dosimeter available that could record dose to such high precision under these circumstances in 3-dimensions. While the gel dosimeter doesn't represent the high density bone of the T12 vertebrae, this shows that when some inhomogeneous regions are introduced, such as lung and patient chest wall, the planning system and delivery systems are capable of achieving a safe outcome.

4. Conclusion

The Mid North Coast Cancer Institute has recently commenced an SBRT paraspinal treatment program. Using VMAT with high doses near to critical serial organs at risk is complex and requires validation using 3-D dosimeters. True 3-D dosimeters are not commercially available, but with the use of an in-house polymer gel dosimetry program this study enables a greater level of confidence in delivery and allows the paraspinal SBRT program to safely advance its treatment options. This study found that the full delivery chain, from end-to-end can safely deliver a high dose fractionation regime to complex and tight margins, ultimately resulting in better patient outcomes.

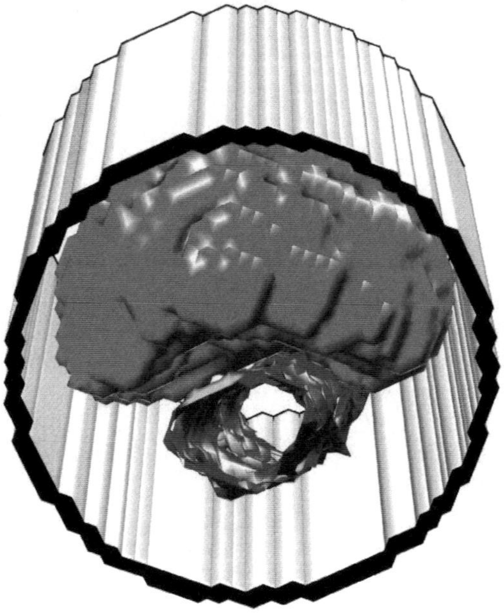

Figure 6. 5Gy isodose sphere (red) with spinal cord (blue) inside the gel filled PET vial.

5. References

[1] Vial *et al* 2006 *Phys. Med. Biol.* **51** 5517-38
[2] Hill B *et al* 2002 *Phys. Med. Biol.* **47** 4233
[3] De Deene Y and Baldock C 2002 *Phys. Med. Biol.* **47** 3117-41
[4] De Deene Y *et al* 2002 *Phys. Med. Biol.* **47** 2459-70
[5] De Deene Y *et al* 2002 *Phys. Med. Biol.* **47** 3441-63
[6] Baldock C *et al* 1998 *Phys. Med. Biol.* **43** 695-702
[7] Baldock C *et al* 2010 *Phys. Med. Biol.* **5** R1-R63
[8] Venning A J *et al* 2005 *Phys. Med. Biol.* **50** 3875-88
[9] Venning A J *et al* 2004 *J. Phys.: Conf. Ser.* **3** 155-8
[10] Baldock C 2009 *J. Phys: Conf. Ser.* **164** 012002
[11] Venning A J *et al* 2005 *Med. Phys.* 32 1047-53
[12] Venning A J *et al* 2019 *J. Phys.: Conf. Ser* **1305** 012003
[13] Venning A J *et al* 2020 *Submitted to IC3DDose* 2020
[14] De Deene Y *et al* 2006 *Phys. Med. Biol.* **51** 653-73
[15] Sahgal A *et al* 2010 *Int. J. Radiation Oncology Biol. Phys.* **77** 548–53
[16] Murry P *et al* 2000 *Austral. Phys. Eng. Sci. Med.* **23** 44-51

Journal of Physics: Conference Series **2167** (2022) 012017 doi:10.1088/1742-6596/2167/1/012017

Study of the use of gel dosimetry in combination with 3D printing phantom for personalized pretreatment QA in radiotherapy

J Colnot[1], S Chiavassa[2], G Delpon[2], C Huet[1]

[1] Institut de Radioprotection et de Sûreté Nucléaire (IRSN), Service de Recherche en Dosimétrie, Laboratoire de Dosimétrie des Rayonnements Ionisants, Fontenay-aux-Roses, France

[2] Institut de Cancérologie de l'Ouest, Département de Physique Médicale, Site Saint-Herblain, France

Email : christelle.huet@irsn.fr

Abstract. In modern radiotherapy, pretreatment patient-specific quality assurance (PSQA) generally consists in delivering the treatment plan to a phantom equipped with a detector and in comparing the measured dose and the dose calculated by the treatment planning system (TPS) in order to detect any gap between both dose distributions. Dosimetric gels have interesting properties for QA. In this work, the use of gel dosimetry together with a patient-based 3D printed phantom for personalized PSQA is investigated. CT images of a patient with a right mesencephalic brain tumor were used to generate a 3D printed phantom. Then it was filled with water and a radiochromic gel jar and irradiated according to the patient intracranial stereotactic plan using a Novalis TrueBeam STX accelerator. Measured dose distributions agree well with the calculated ones. Regarding 3D gamma-index (1 mm – 2%) estimated within the central 85% of the jar volume, 96.3% of points pass the test. In addition, 86.5% of the points pass the local 2D 3 mm-3% gamma-index. Results are promising but further work is needed to improve the protocol and investigate the possibility to extend it to end-to-end tests.

1. Introduction

Modern radiotherapy techniques such as intensity modulated radiation therapy, volumetric modulated arc therapy or stereotactic radiotherapy are more and more complex. Thanks to the use of either multiple beams or arcs with simultaneously gantry rotation, MLC shape adjustment and dose rate modulation, highly conformal doses to the tumor volume is delivered while preserving close organs at risk. It is a requirement that medical physicists conduct pretreatment QA measurements to detect any gap between calculated and delivered doses. This generally consists in delivering the plan to a phantom and comparing both measured and calculated dose distributions. With advanced RT techniques, due to the complex shape of dose distributions, QA is challenging and accurate and spatially precise dosimeters are needed in combination with new PSQA procedures. Dosimetric gels have interesting properties for QA [1]: in addition to the ones mentioned above they are water equivalent and 3-D dose distributions can be obtained in a single irradiation. Regarding phantoms, commercially available phantoms are not always anatomically correct. However with the arrival and still maturing technology of three-dimensional printing, it is now possible to construct personalized phantoms from the CT images of the patients [2].

This study describes the feasibility of using 3D-printed patient specific phantoms in combination with a radiochromic gel in order to perform pretreatment patient-specific quality controls in stereotactic radiotherapy.

2. Material and methods

2.1. Phantom construction and gel manufacture

A patient with a right mesencephalic brain tumor was selected. The first step consisted in creating a numerical phantom from the CT images of the patient. Then this numerical model was converted into a mesh using STL format. Finally the phantom was 3D printed at scale 1:1 using an Ultimaker 3 extended printer which is based on fused deposition modeling. The phantom consisted in a 1.2mm thick wall made of PET including a place to insert a radiochromic gel jar (Figure 1). The nose was printed separately to be used as a plug [3].

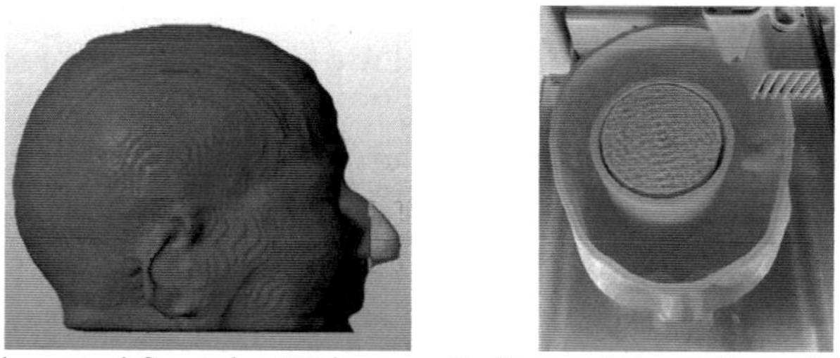

Figure 1. 3D model created from the CT images (left) and 3D printing of the model with a gel insert (right)

An in-house protocol was used for the preparation and the calibration of the radiochromic TruView™ gels [4]. Two jars of 400 ml from the same batch were used: one for the calibration and another dedicated to the measurement in the phantom. Before irradiation, 512 projections of the gel jars were acquired at 632 nm using the commercial Vista™ Optical-CT scanner.

2.2. Treatment planning

CT images of the phantom equipped with the stereotactic mask and including the gel jar were acquired and imported in the iPlan TPS (BrainLab) to recalculate the intracranial stereotactic treament plan (Figure 2). The original plan consisted in 3 fractions of 11 Gy, it has been adapted in order to deliver 3 Gy to the target volume which is the optimal dose for the TruView™ gel. The ballistic was composed of 6 dynamic arc beams, the total number of delivered MU was 451.

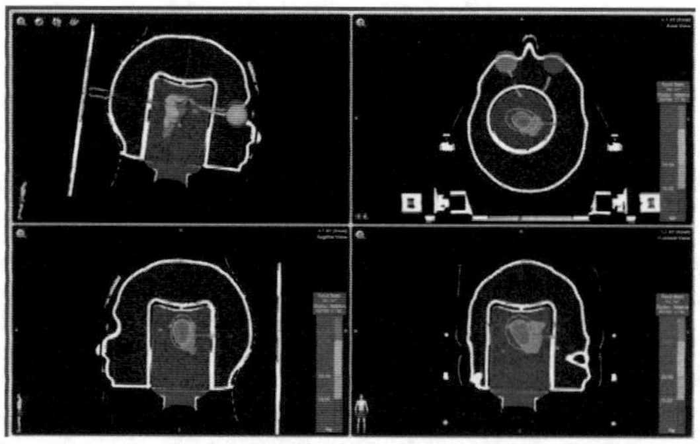

Figure 2. Axial, sagittal and coronal views of the iPlan computed dose distribution created using the 3D-printed phantom

Journal of Physics: Conference Series **2167** (2022) 012017 doi:10.1088/1742-6596/2167/1/012017

2.3. Irradiation
For the calibration, the jar was placed vertically in a water tank without the lid and irradiated with a 12 MeV electron beam and a 6.5 cm×6.5 cm field. The maximum dose delivered was 4 Gy.

The phantom with the gel jar was placed on the treatment table with the stereotactic mask. Then, kV images using the ExacTrac system were acquired in order to allow a precise repositioning of the phantom. Finally, one treatment fraction of 3 Gy was delivered on the phantom using a Novalis TrueBeam STX accelerator (Varian Medical System).

2.4. Gel reading and data analysis
Gels were optically read 75 minutes after irradiation. Using the VistaRecon™ software (Modus Medical Devices Inc.), reconstructions were made with a voxel resolution of 0.5 mm × 0.5 mm × 0.5 mm. Then the data were processed using Matlab (vR2013b, The MathWorks, Inc.). For the calibration jar, the central-axis changes in optical attenuation with depth of gel dosimeters were fit against the central-axis depth dose extracted from the TPS library. For the jar in the phantom, the quantitative comparisons included dose distributions and 2D and 3D gamma index.

3. Results and discussion
As observed in Figure 3, the measured dose distribution is in agreement with the calculated one (Figure 3). The shape and the dose levels are similar. Regarding 3D gamma-index (1 mm – 2%) estimated within the central 85% of the jar volume, 96.3% of points pass the test (Figure 4). For a slice located at the center of the gel dosimeter, the local 2D 3 mm-3% gamma-index evaluated within the central 85% of the cylinder diameter reaches 86.5% of points passing the criteria.

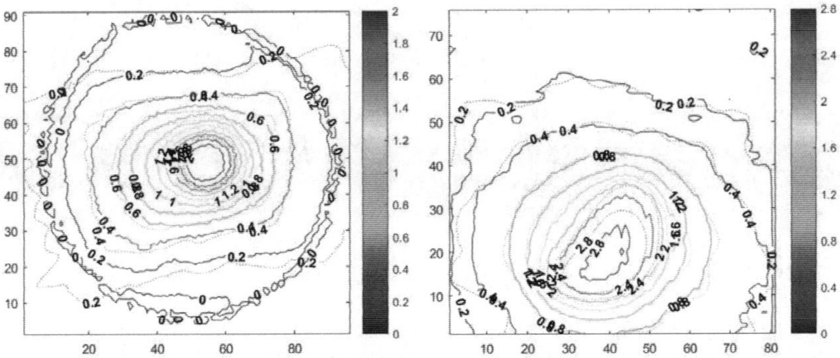

Figure 3. 2D dose distributions obtained with the gel (dotted lines) and with the TPS (solid line) in an axial plane (left) and a coronal plane (right) across PTV

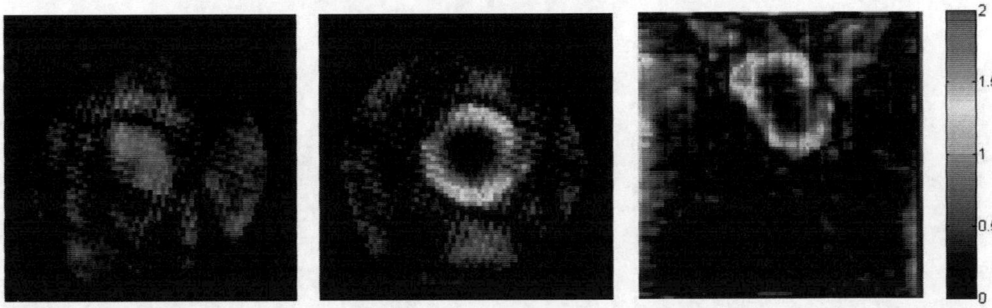

Figure 4. 3D global gamma analysis (1 mm – 2%) in two axial planes (left and middle) and in a coronal plane (right) at PTV level

However, it can be observed that some discrepancies are observed in the strong dose gradient around the target volume (Figure 4 and Figure 5). The gel underestimates the dose in that region. This could be due to the diffusion of ferric ions through the gel which causes the degradation of the spatial and dosimetric information with time. The protocol has been developed to limit this diffusion but these measurements haven't been performed in control laboratory conditions as usual in particular in terms of temperature. The TruView^{TM} gels are temperature-sensitive and their response may have been slightly impacted. Moreover, this underestimation could also come from reading artefacts with the optical scanner. As a matter of fact, the tank needs to be filled and degassed the day before. However due to time constraints, it has only been done the day of measurements.

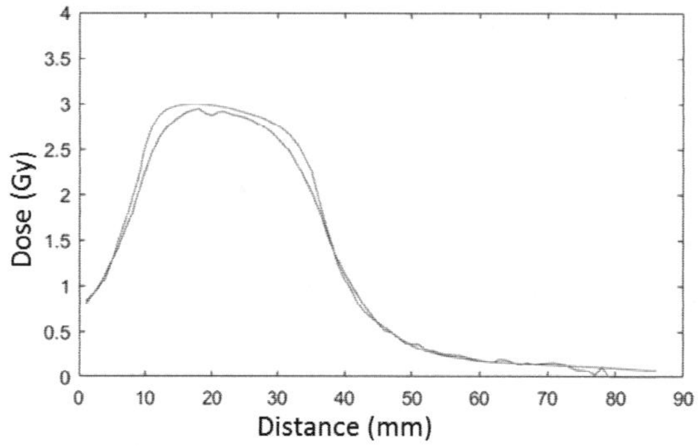

Figure 5. Dose profiles of the calculated and measured axial dose distribution at PTV centre along the head-foot direction

4. Conclusion

Gel dosimetry in combination with patient-based 3D-printed phantoms is a promising technique for personalized PSQA in stereotactic radiotherapy. However, further studies are needed to improve the protocol. Future work will also investigate the possibility to extend the method to end-to-end tests.

5. References

[1] Babic S, Battista J and Jordan K. 2008 *Int J Radiat Oncol Biol Phys* **70** 1281-1291
[2] Ehler E D, Barney B M, Higgins P D and Dusenbery K E. 2014 *Phys Med Biol* **59** 5763-5773
[3] Leborgne F. 2018 *Master thesis*
[4] Colnot J, Huet C, Gschwind R and Clairand I. 2018 *Phys Med* **52** 154-164

RSC: A 3D printed eyeball phantom for Sr-90 dosimetry measurements

S K Maxwell[1], S B Crowe[1,2,3], E M Simpson-Page[1], N Cassim[1] and T Kairn[1,2]
[1]Royal Brisbane and Women's Hospital, Brisbane, QLD, Australia
[2]Queensland University of Technology, Brisbane, QLD, Australia
[3]Herston Biofabrication Institute, Herston, QLD, Australia

Email: Sarah.Maxwell@health.qld.gov.au

Abstract. Strontium 90 (Sr-90) has been commonly used in the radiation treatment of pterygia of the eye. A radioactive plaque is placed in an applicator onto the surface of the eyeball for a specific length of time to achieve a desired dose. Dose is usually calculated using source activity and decay, as well as the distance from it to the surface of the eye. However, this assumes a flat eye surface. This investigation used 3D printing to produce an anthropomorphic eyeball phantom on which to perform dosimetry measurements for two different applicator sizes. These doses were compared to planar geometry measurements and a dose difference found. While planar geometry measurements are useful for routine quality assurance, measurements of the effects of the curved surface on dose calculations can provide valuable clinical information.

1. Introduction

Strontium 90 (Sr-90) is a valuable tool for radiation treatment of pterygia due to the short range of beta particles it emits. Pterygia is a benign growth that occurs on the surface of the eye. Commonly used post-surgery, the treatment is an effective method to help prevent recurrence of the condition. It involves the use of a radioactive plaque placed in an applicator on the eyeball for a specific length of time to achieve the desired dose [1]. The dose can be calculated using simple estimates based on source activity and the distance from it to the surface of the eye, dependent on the applicator. Sr-90 has a half thickness of 1.5 mm, meaning after 1.5 mm penetration through water, the dose rate is attenuated by 50%, making it ideal for use with surface therapy [2].

A number of studies exist on the calibration of Sr-90 sources using a variety of dosimeters [3-5]. The ICRU report 72 provides guidelines for the dosimetry of beta rays for brachytherapy with sealed sources [6]. The 2015 paper by Laoues et al [7] simulated a Sr-90 treatment using Monte Carlo calculations and compared this to physical measurements performed with film. A good agreement was found between the film measurements and the Monte Carlo results; however, these doses were higher than that provided by the manufacturer. No investigations found in the literature however, have performed measurements on an anthropomorphic eyeball model that allows the curvature of the surface of the eye to be taken into account.

This study therefore used medical imaging data to design an eyeball with anatomically realistic geometry and density, which was fabricated using 3D printing and used to perform film dosimetry measurements of Sr-90 treatments using two applicators of different aperture sizes.

2. Method

2.1. Fabricating the eyeball phantom
MRI datasets of 13 healthy volunteers were used to take dimensions of 26 eyeballs. The collection of this MRI data for use in radiotherapy phantom design was approved by the Metro North Hospital and Health Service Human Research Ethics Committee. Each eyeball was measured in the transverse, sagittal and axial planes. The dimensions for each lens were also recorded. These dimensions were used to obtain an average eyeball size.

The HU values of the different structures of the eye were taken from CT datasets. An average HU value of approximately +1 was found for the vitreous humour of the eyeball and a value of +33 was obtained for that of the lens. Due to the relatively small difference between the vitreous humour and lens densities, it was decided not to include the lens in the print as the uncertainty in the density of the printed material outweighed the difference.

An average eyeball model was designed using Tinkercad (Autodesk Inc., San Rafael, USA). Initially, an eyeball was contoured in MIM Maestro ver. 6.7.6 (MIM software Inc, Cleveland, USA) on one of the MRI datasets and an STL of the volume exported. This was then scaled to match the average dimensions calculated. However, it was found that the voxelisation from the CT images meant the model was not smooth or an accurate shape. Instead, the average dimensions were used to scale and shape a simple sphere in Tinkercad, to achieve the required eyeball shape. This was then sliced into quarters for film planes and 3D printed in PLA with 86% infill using a Raise 3D (Raise 3D Technologies, Irvine, USA) Pro 2 FDM 3D printer, in polylactic acid (PLA). A holder to support the eyeball and provide scatter was also printed using PLA. The eyeball, in position in the holder, was CT-scanned and analysed for any inconsistencies in the print using in-house 3D print QA software [8].

2.2. Dosimetry measurements
Radiochromic film was used for the dosimetry of the two Sr-90 applicators, with respective aperture diameters of 1.6 and 1.0 cm, while the Sr-90 source had an active diameter of 1.1 cm. Film was prepared as per recommended methods, including pre-irradiation scans [9]. Two calibrations were performed using a 6 MV beam and a 300 kVp beam with ten pieces of film irradiated with doses of 0 – 400 cGy for each. The effective energies were interpolated between to derive a calibration curve for Sr-90.

Two measurement setups were used for each applicator, one with the film in the vertical cross section of the 3D printed eyeball model (Figure 1) and one with film on the surface of a stack of virtual water for a planar geometry (Figure 2). Two pieces of film were irradiated for each applicator, with each film being irradiated for 1 minute.

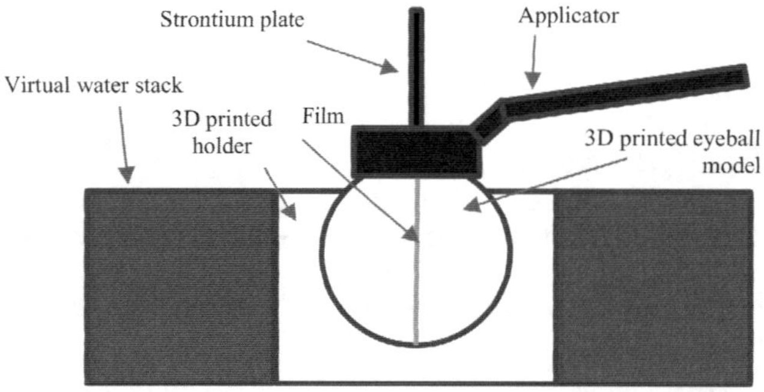

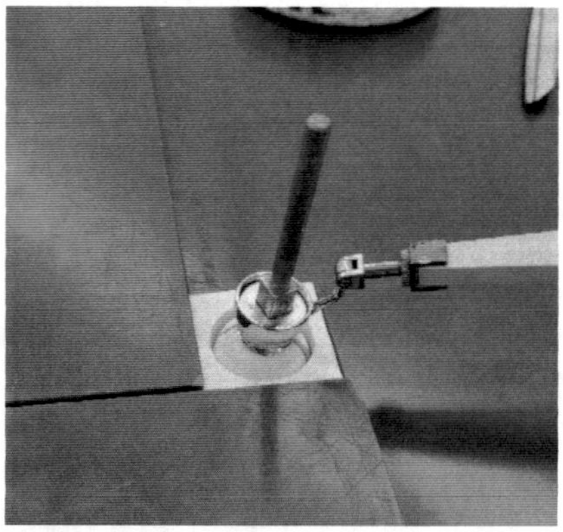

Figure 1. Set up of eyeball phantom and film with Sr-90 source and applicator

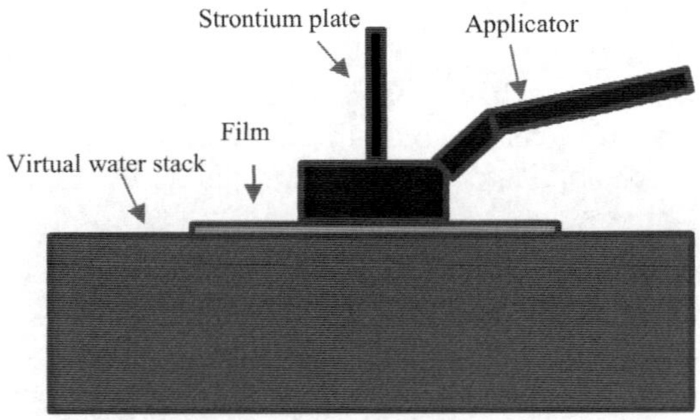

Figure 2. Planar geometry set up pf film on virtual water with Sr-90 source and applicator

Approximately 48 hours after irradiation, each piece of irradiated film was scanned with the same setup as the pre-irradiation scans to allow pixel values to be converted to dose using the derived calibration relationship, according to an established film analysis method [9, 10].

Dose images were calculated, and the dose distributions were analysed for each applicator and the planar geometry measurements. Multiple line plots across the vertical plane were used to measure the dose in the vertical plane, perpendicular to the source, for each applicator film measurement.

As the edges of the film were observed to be effected by delamination despite careful cutting with scissors [11], doses measured in the eyeball phantom for each applicator were compared at a depth of 0.15 cm, where the dose was unaffected by delamination on all films, rather than at the surface, which would have been ideal.

3. Results

3.1 The eyeball phantom

The average diameters of the eyeballs analysed from the MRI datasets and used for the eyeball phantom design were 24.51 ± 0.02 mm, 24.63 ± 0.02 mm and 25.17 ± 0.02 mm in the sagittal, transverse and vertical directions, respectively. These dimensions agree with the mean eyeball diameters given in ICRP "Report of the task group on reference man" [11]. The printed model and the QA results can be seen in Figure 3.

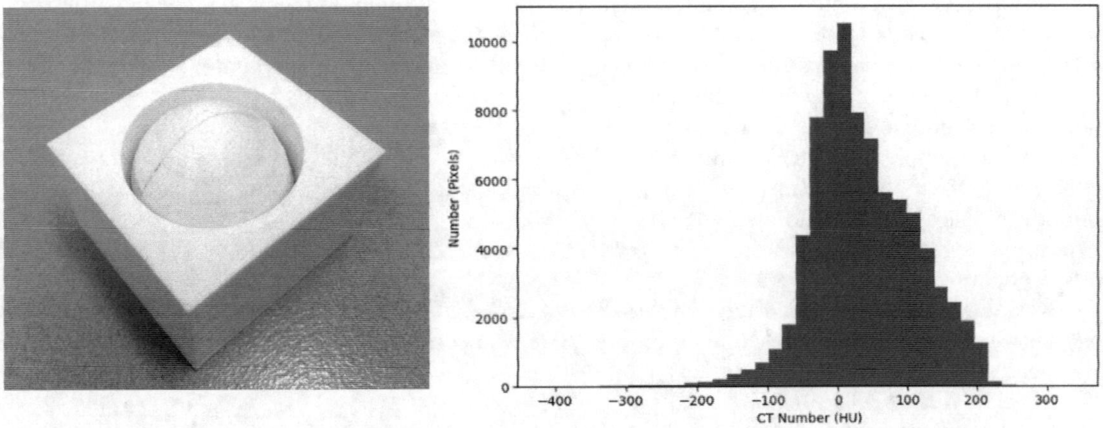

Figure 3. The 3D printed eyeball phantom in its holder and the QA analysis results showing HU histogram.

Journal of Physics: Conference Series **2167** (2022) 012018 doi:10.1088/1742-6596/2167/1/012018

3.2 Dosimetry measurements

The dose to the surface of the eye was calculated using the calibration certificate of the source, the decay of Sr-90 and a calculated standoff factor (account for the distance between the source and the eye due to the applicator design) for each applicator. These calculated doses, as well as the measured doses for each setup are presented in Table 1.

Table 1. Calculated and measured doses, for each applicator

Applicator diameter	Applicator standoff	Calculated dose (Gy/min)	Measured planar dose at surface (Gy/min)	Measured eyeball dose at 0.15 cm depth (Gy/min)
1.6 cm	0.8 cm	1.9	1.9 ± 0.2	1.6 ± 0.1
1.0 cm	0.6 cm	3.4	3.7 ± 0.4	2.8 ± 0.1

4. Discussion

The results shown in Table 1 indicate that the measured dose from the planar set up was in agreement with the calculated dose. This result is unsurprising, given that the dose calculation assumes a flat eyeball. While the literature suggests that at a depth of 0.15 cm the dose should be 50% of the maximum, the measurements performed on the eyeball phantom give a dose approximately 80% of the surface dose from the planar setup. This would be due to the curvature of the eyeball surface, which brings the treated tissue closer to the source than the applicator standoff distance, and is a more realistic approximate of what a patient would receive at this depth during a treatment. It would have been ideal to measure the dose at the surface of the eyeball phantom, but delamination is a common issue encountered in film dosimetry [12].

It is therefore advisable to use planar measurements (such as shown in Figure 2) for any routine quality assurance of the Sr-90 source, but as there is a difference between measurements performed on a flat set up and those done on an anthropomorphic phantom, the 3D printing of an eyeball phantom can help give a more realistic indication of the dose received by a patient undergoing this treatment. The methods used in this work can also be used by other radiotherapy centres performing ocular treatments with Sr-90, or other sources, to print 3D models and perform dosimetry with local applicator systems, to test the assumptions underlying treatment dose calculations.

5. Conclusion

A 3D printed eyeball phantom was printed and used with radiochromic film to investigate the difference in dose between planar geometry and a more realistic dose the patient would receive when being treated with Sr-90 for pterygia. The planar geometry measurements agreed with simple dose calculations for the source used, while the eyeball measurements at a depth of 0.15 cm gave a higher dose relative to the planar surface dose at this depth due to the curved eyeball surface.

While simple, planar phantoms are useful for routine quality assurance, 3D printed anthropomorphic phantoms can provide important additional information regarding the doses delivered to more complex human anatomy.

6. References

[1] Kearsley JH, Fitchew RS, Taylor RGS 1988 *Int. J. Radiat. Oncol. Biol. Phys.* **14** 435-443
[2] Kirwan JF, Constable PH, Murdoch IE *et al* 2003 *Nature* **17** 207-215
[3] Soares CG 1992 *Phys. Med. Biol.* **37** 1005-1007
[4] Soares CG *et al* 2001 *Med. Phys.* **28** 7:1373-1384
[5] Siddle D, Langmack K 1999 *Phys. Med. Biol.* **44** 7
[6] ICRU Report No. 72 2004 *Journal of ICRU* 4:0-175
[7] Laoes M, Khelifi R, Moussa AS 2015 *J. Phys.: Conf. Ser.* **573** 012072

[8] Charles PH, Kairn T, Crowe SB 2020 *Phys. Eng. Sci. Med.* **43** 436-437
[9] Aland T, Kairn T, Kenny J 2011 *Australas Phys Eng Sci Med.* **34** (2):251-60
[10] Spelleken E, Crowe SB, Sutherland B *et al* 2018 *Australas. Phys. Eng. Sci. Med.* **41** 117-128
[11] ICRP 1975 *ICRP publication 23.* Pergamon Press, Oxford
[12] Zolfaghari S, Francis KE, Kairn T *et al* 2017 *Australas. Phys. Eng. Sci. Med.* **40** 449-453

Journal of Physics: Conference Series **2167** (2022) 012019 doi:10.1088/1742-6596/2167/1/012019

Effects of radical scavengers on nanocomposite Fricke gel for heavy ion beam irradiation

T. Maeyama[1,2], N. Fukunishi[2], K. L. Ishikawa[2,3], K. Fukasaku[4,5] and S. Fukuda[6]
[1]Kitasato University/Department of Chemistry, School of Science, Kanagawa, Japan
[2]RIKEN/Nishina Center for Accelerator-Based Science, Saitama, Japan
[3]The University of Tokyo/Department of Nuclear Engineering and Management, Graduate School of Engineering, Tokyo, Japan
[4]Department of Neurosurgery, Niiza-Shiki Chuo General Hospital, Saitama, Japan
[5]RIKEN/Center for Advanced Photonics, Image Processing Research Team
[6]National Institutes for Quantum and Radiological Science and Technology/QST Hospital, Chiba, Japan

Email: maeyama@kitasato-u.ac.jp

Abstract. NC-FG (nanocomposite Fricke gel) dosimeter is a 3D dosimeter for heavy ion beam without LET dependence. In this study, we evaluate the effects of silver perchlorate, a radical scavenger, on NC-FG. We find that radiological properties of NC-FG are changed by small amounts of silver perchlorate. Especially, dose response at high LET enhanced.

1. Introduction

Three-dimensional (3D) dosimetry of heavy ion beam irradiation is required in particle therapy using carbon ion beam. On the other hand, the practical application of a solid-state or gel-state dosimeter to heavy ion beam 3D dosimetry is complicated by the degradation in sensitivity as a function of linear energy transfer (LET) [1-3]. Although Fricke gel [4-8] dosimeter also shows LET dependent dose responses [9,10], the Fricke gel prepared with nano-clay under deaerated conditions (NC-FG: nanocomposite Fricke gel) shows LET-independent dose responses, and this is the only 3D dosimeter for heavy ion beam [10-12]. NC-FG show the reaction scheme with water radiolysis products as shown

$$\boxed{\text{Clay}\ \overset{-}{\underset{-}{}}}\,\text{Fe}^{2+} + \text{OH} \rightarrow \boxed{\text{Clay}\ \overset{-}{\underset{-}{}}}\,\text{Fe}^{3+} + \text{OH}^-$$

$$\boxed{\text{Clay}\ \overset{-}{\underset{-}{}}}\,\text{Fe}^{3+} + e_{aq}^- \rightarrow \boxed{\text{Clay}\ \overset{-}{\underset{-}{}}}\,\text{Fe}^{2+}$$

$$\boxed{\text{Clay}\ \overset{-}{\underset{-}{}}}\,2\text{Fe}^{2+} + \text{H}_2\text{O}_2 \rightarrow \boxed{\text{Clay}\ \overset{-}{\underset{-}{}}}\,2\text{Fe}^{3+} + 2\text{OH}^-$$

$$\text{G}(\text{Fe}^{3+})_{\text{Clay}} \cong \text{G}(\text{OH}) + 2\text{G}(\text{H}_2\text{O}_2) - \text{G}(\text{e}_{\text{aq}}^-)$$

Figure 1. Reaction scheme of NC-FG

Journal of Physics: Conference Series **2167** (2022) 012019 doi:10.1088/1742-6596/2167/1/012019

in figure 1. The reaction part of reduction of Fe^{3+} by hydrated electron is major factor for LET-independent dose responses [11]. In this paper, we investigate the effects of silver perchlorate for a radical scavenger on NC-FG for heavy ion beam irradiation, and we try to control of LET-dependence of dose response of NC-FG.

2. Material and methods

The NC-FG prepared with 2 wt% nano clay (Laponite XLG) and 1 mM ferrous sulfate ammonium base on reference [11]. Ag^+ ion was added for radical scavenger. The addition of silver ion Ag^+ (40% Silver Perchlorate Solution) was done in nano-clay dispersion solution while bubbling through H_2 gas and mixing by magnetic stirrer (Nissin magnetic stirrer Model SW-R300) for 1 hour. After color change colorless to brown of nano-clay dispersion plus silver ion solution (0.1, 0.5 or 1 mM), we added the ferric sulfate ammonium stock solution, and mixing using same magnetic stirrer for 1 hour while bubbling through H_2 gas. The obtained almost black solution after stirring was sealed into Pyrex glass tubes under a pure N_2 atmosphere in a glove box. Details of the chemical compositions of NC-FGs are summarized in Table 1. The prepared Pyrex glass tubes containing NC-FG were irradiated from the bottom with A $^{12}C^{6+}$ beam at 290 MeV/u supplied from the Biological Irradiation Port of Heavy Ion Medical Accelerator in Chiba (HIMAC) at the National Institute of Radiological Sciences (NIRS). Detailed irradiation condition is same with report [11]. A 1.5 T MRI scanner (Intera Achieva 1.5T HP Nova Dual Gradient, Philips Medical Systems, Best, The Netherlands) was used for the analysis of the NC-FG samples. The longitudinal MR relaxation rate ($R_1 = 1/T_1$) of the NC-FG samples was evaluated using a turbo mixed sequence. The conditions of the T_1 measurements were as follows: TR = 2260 ms; TE_1 = 19 ms; TE_2 = 100 ms; TI = 500 ms; ETL = 6; pixel spacing = 0.8 mm.

Table 1. Chemical composition of NC-FG

	ammonium iron(II) sulfate hexahydrate (mM)	LaponiteXLG (w/w%)	silver(I) perchlorate (mM)	Bubbling
1	1	2	0	Deaeration
2	1	2	0.1	H_2
3	1	2	0.5	H_2
4	1	2	1	H_2

3. Result

In the fig 2, Monotonically increase of R_1 with dose can be observed in the experiments results of addition of Ag^+ ions. On the other hand, the radiation sensitivity of NC-FG decreases with increase in Ag^+ concentration. Ag^+ ions have high reactivity with OH radical and also with hydrated electron [13]. Thus, it is considered that the oxidation of Fe^{2+} was suppressed by the addition of Ag^+ ions. In the figure 3, Depth-δR_1 distributions was shown, and the values of δR_1 plotted as the normalized value at entrance surface. we can confirm that peak enhancement with increase in concertation Ag^+ from the relative radiation sensitivity plot. It is means that the contribution of reduction reaction of Fe^{3+} is relatively increase by addition of Ag^+ ion. If NC-FG is not prepared by using hydrogen bubbling before mixing Ag^+ and Fe^{2+}, rapidly oxidation before irradiation experiment was observed. The understanding of these interesting phenomena needs to the other analyzing method such as optical spectroscopy shown in literature [14]. In addition, we must also investigate dose rate effects and change over time. Enhancement of sensitivity with increase in LET value is not unique in this dosimeter but also known in $Fe^{2+} + Cu^{2+}$ aqueous chemical dosimeter [15]. We speculate that NC-FG added Ag^+ radical scavenger also has same reaction system with $Fe^{2+} + Cu^{2+}$ aqueous chemical dosimeter.

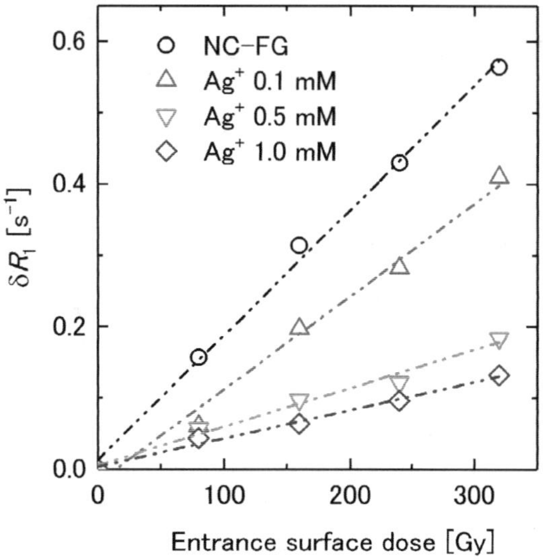

Figure 2. The δR_1 value of the NC-FG after irradiation of $^{12}C^{6+}$ 290MeV/u. The symbols indicate measured values (MRI) near the entrance surface. The dashed line indicates liner fitting.

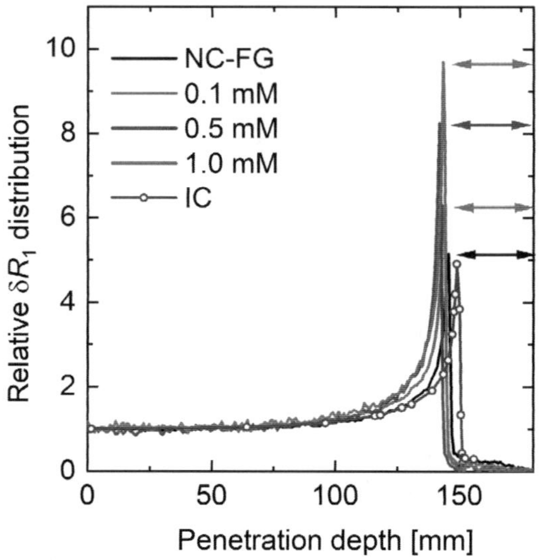

Figure 3. The relative δR_1 distribution of NC-FG after irradiation of $^{12}C^{6+}$ 290 MeV/u. The symbol-line show dose measured using IC (ionization chamber).

4. Conclusion

We reported that the further development of improvement radiological property of NC-FG as a function of LET. LET effect was changed using small amount of radical scavenger of Ag^+ ion. While LET effect enhancement by addition of Ag^+ ion is only phenomenological understanding, this study is clearly show NC-FG can control LET effects. It is maybe useful for development such as biological detector that has similar LET effects in the future. On the other hand, that sensitivity decrees as shown in fig 2. For more detailed characterization, it is necessary to improve the sensitivity or change the measurement system. For example, it is useful to investigate the application of a gel dosimeter using the most sensitive fluorescent probe [16-20].

5. Acknowledgments

This research was supported by MEXT (JSPS) KAKENHI (26820412, 17K07013). A part of this study was performed as the Research Project with Heavy Ions at NIRS-HIMAC.

6. References

[1] Karger C P, Jakel O, Palmans H *et al* 2010 *Phys. Med. Biol.* **55** R193-R234
[2] Maeyama T, Fukunishi N, Ishikawa K L *et al* 2015 *Radiat. Phys. Chem.* **107** 7-11
[3] Furuta T, Maeyama T, Ishikawa K L *et al* 2015 *Phys. Med. Biol.* **60** 6531-46
[4] Gallo S et al. 2018 *Sens. actuators. B Chem.* **272** 618-25
[5] Gallo S, Artuso E, Brambilla M G *et al* 2019 *J. Phys. D: Appl. Phys.* **52** 225601
[6] Smith S T, Boase N R B, Masters K S *et al* 2019 *Phys. Med. Biol.* **64** 205017
[7] Babu S E S, Peace B.S T, Rafic K M *et al* 2019 *Radiat. Phys. Chem.* **156** 300-06
[8] Hill B *et al* 2002 *Phys. Med. Biol. 47* 4233-46
[9] Buglewicz D J, Banks A B, Hirakawa H *et al* A 2019 *Scientific Reports* **9** 6157
[10] Maeyama T, Fukunishi N, Ishikawa K L *et al* 2014 *Radiat. Phys. Chem.* **96** 92-96

[11] Maeyama T, Fukunishi N, Ishikawa K L *et al* 2017 *J. Phys. Chem. B.* **121** 4238-46
[12] Maeyama T, Fukunishi N, Ishikawa K L *et al* 2016 *J. Radiat. Res.* **57** 318-24
[13] NIST Chemical Kinetics Database N S R D, Version 7.0 (Web Version), Release 1.6. 8, Data version 2015.12, National.
[14] Islam K T, Dempsey J F, Ranade M K *et al* 2003 *Med. Phys.* **30** 2159-68
[15] Bjergbakke E, Sehested K, Rasmussen O L *et al* 1976 *Radiat.Research* **66** 433-42
[16] Watanabe Y, et al. 2020 *Phys. Med. Biol.* **65** 175008
[17] Mochizuki A, Maeyama T, Watanabe Y *et al* 2020 *RSC Advances* **10** 28798-806
[18] Maeyama T, Mochizuki A, Takanashi T 2019 *J. Phys. Conf. Ser.* **1305** 12045
[19] Maeyama T, Kato A, Mochizuki A *et al* 2019 *Sens. Actuators A Phys.* **298** 111435
[20] Maeyama T, Hase S 2018 *Radiat. Phys. Chem.* **151** 42-46

Journal of Physics: Conference Series **2167** (2022) 012020 doi:10.1088/1742-6596/2167/1/012020

RSC: Gel dosimetry as a tool for clinical implementation of image-guided radiotherapy

A Elter[1,2,4], S Dorsch[1,2,4], M Marot[1,4], C Gillmann[1,4], W Johnen[1,4], A Runz[1,4], C K Spindeldreier[3,4], S Klüter[3,4], C P Karger[1,4] and P Mann[1,4,5]
[1] Department of Medical Physics in Radiation Oncology, German Cancer Research Center (DKFZ), INF 280, 69120 Heidelberg, Germany
[2] Faculty of Physics and Astronomy, University of Heidelberg, Heidelberg, Germany
[3] Department of Radiation Oncology, University Hospital Heidelberg, INF 400, 69120 Heidelberg, Germany
[4] National Center for Radiation Research in Oncology (NCRO), Heidelberg Institute for Radiation Oncology (HIRO), Heidelberg, Germany
[5] HQ-Imaging GmbH, Heidelberg, Germany

E-Mail: a.elter@dkfz.de

Abstract. The implementation of new image-guided radiotherapy (IGRT) treatment techniques requires the development of new quality assurance (QA) methods including geometric and dosimetric validation of the applied dose in 3D. Polymer gels (PG) provide a promising tool to perform such tests. However, to be used in a large variety of clinical applications, the PG must be flexibly applicable. In this work, we present a variety of phantoms used in clinical routine to perform both hardware and workflow tests in IGRT. This includes the validation of isocenter accuracy in magnetic resonance (MR)-guided RT (MRgRT) and end-to-end tests of online adaptive treatment techniques for inter- and intra-fraction motion management in IGRT. The phantoms are equipped with one or more PG containers of different materials including 3D printed containers to allow for 3D dosimetry in arbitrarily shaped structures. The proposed measurement techniques and phantoms provide a flexible application and show a clear benefit of PG for 3D dosimetry in combination with end-to-end tests in many clinical QA applications.

1. Introduction

New radiotherapy (RT) treatment techniques, such as image-guided RT (IGRT), applied in clinical routine increase tumor-conformity while sparing surrounding organs at risk (OAR). However, the increasing complexity of such techniques requires the development of new quality assurance (QA) procedures ensuring correct dose delivery. Especially for new hybrid machines combining magnetic resonance imaging (MRI) and conventional RT (so-called MR-Linacs), the implementation of new QA protocols is of great interest. These protocols range from single component testing to the validation of new motion management and adaptive treatment strategies (such as gating and daily plan adaption). However, the complexity of such tests requires dedicated detectors allowing to test (i) geometric and (ii) dosimetric accuracy of the 3D dose distribution in (iii) phantoms, which are visible in MRI. One promising method meeting requirements (i)-(iii) are so-called polymer gels (PG) [1]. To be used in a larger variety of clinical QA procedures, the PG applications should be as flexible as possible including the use of arbitrarily shaped gel containers to simulate anthropomorphic structures. Furthermore, a geometrical evaluation of the gel directly after irradiation instead of after the long stabilization times of the gel's polymerization processes of up to 48h [2] is of high importance for the evaluation of geometrical parameters. In this paper, we present a set of dedicated phantoms that can be equipped with arbitrary 3D printed gel containers [3] and that are used at the German Cancer Research Center (DKFZ) and University Hospital in Heidelberg, Germany, for a variety of clinical QA procedures of advanced treatment techniques. The experimental PG workflow as validated in [2] was applied in this work.

2. Polymer gel

2.1. Fabrication
As PG, the PAGAT (PolyAcrylamide Gelatin gel fabricated at ATmospheric conditions) was used and in-house produced following the protocol described elsewhere [2].

2.2. Gel containers
We used three different types of gel containers: (i) borosilicate glass flasks and (ii) Barex^{TM} (VELOX GmbH, Hamburg, Germany) containers, which are two well established materials for the use with PG, but which are also limited in size and shape. Hence, we have developed a method to (iii) 3D print gel containers in arbitrary shapes using the VeroClear^{TM} printing material of the Objet500 Connex 3 3D printer (Stratasys, Eden Prairie, USA) [3].

2.3. Evaluation
Gel evaluation was performed either (i) purely geometrically directly after irradiation on a 0.35T MR-Linac (ViewRay Inc., Ohio, USA) or (ii) dosimetrically 48h after irradiation on a diagnostic 3T Prisma^{fit} (Siemens Healthineers, Erlangen, Germany). For this, a turbospin- or multispin-echo sequence with 1/32 equidistant echoes acquired with an isotropic image resolution of 1 mm^3 was used for (i)/(ii). For quantitative dose comparison, a 3D γ evaluation with a passing criterion of 3%/3mm (dose difference/ distance-to-agreement) was employed, taking only dose levels larger than 10% of the maximum dose into account.

3. Clinical applications
A schematic overview of the phantoms presented in this paper is shown in Figure 1. Each phantom is equipped with one or more PG containers to either perform 3D geometric and/or dosimetric measurements in different scenarios.

3.1. 3D isocenter accuracy measurement in MRgRT
To test the alignment of imaging and irradiation isocenter in MRgRT, we used an in-house developed QA phantom [4]. The phantom is equipped with a spherical PG-filled glass flask of 8cm diameter (Fig. 1a). The phantom was aligned with the imaging isocenter of the MRI using dedicated fiducials (Fig. 1b, red marking) while the irradiation isocenter was measured by a star shot irradiation of the PG container. The PG was then geometrically evaluated in 3D using a T_2-weighted turbo spin echo sequence directly after irradiation without moving the phantom to determine the radius of the irradiation isocircle (IC_r) and the distance between the imaging and irradiation isocenter (IC_d). Results showed a mean IC_r/ IC_d of (0.4±0.1)mm/ (0.4±0.6)mm, which were well within the tolerance levels of 0.5mm and 1mm, respectively [4].

3.2. End-to-end tests of online adaptive IGRT procedures

3.2.1. Inter-fractional motion. With the introduction of new IGRT devices, the online adaption of treatment plans in case of inter-fractional anatomy changes became feasible. To validate these techniques, we used two deformable phantoms that can be equipped with 3D printed and PG-filled structures to validate the dose delivery.

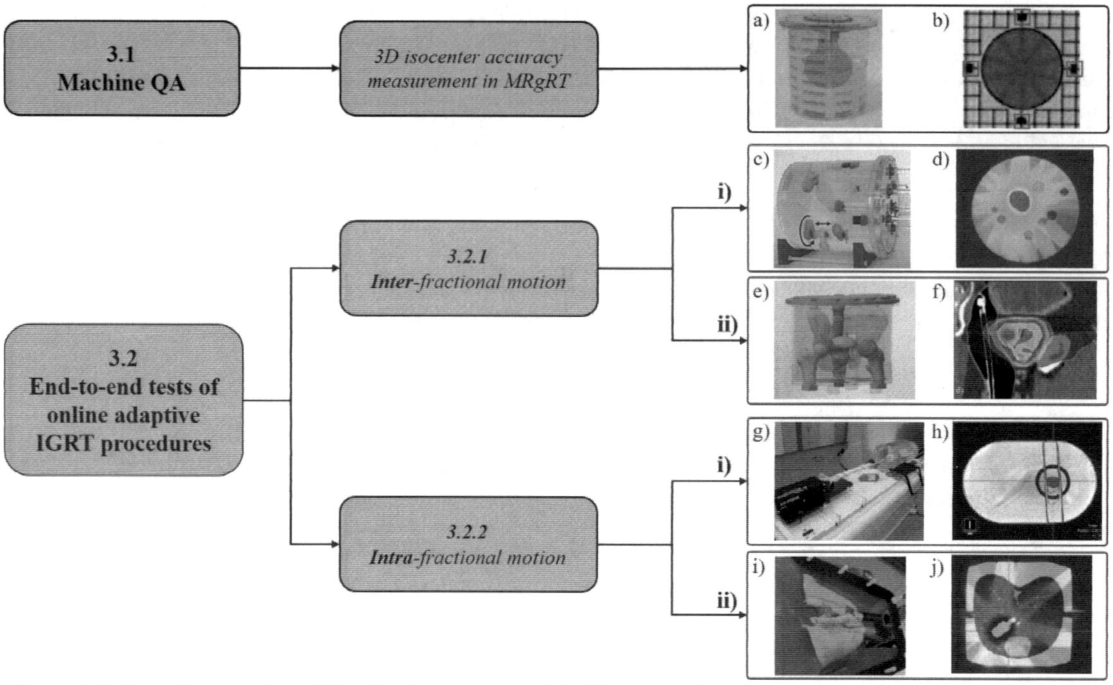

Figure 1. Schematic overview of the phantoms presented in this paper: a) a picture of the phantom for 3D isocenter accuracy measurements in MRgRT including the gel container (green) and b) the geometric PG evaluation including fiducials (red) for a correct positioning in the imaging isocenter. Pictures of the phantoms used for end-to-end tests of online adaptive IGRT procedures are given in c), e), g), and i) together with the respective treatment plan in d), f), h), and j).

i. AQUARIUM. To validate the online plan adaption process in MRgRT the AQUARIUM (Anthorpomorphic QUality AssuRance phantom to study Interfractional Uncertainties in MRgRT) [5] was equipped with two PG-filled structures (OAR, tumor) as well as structures filled with anthropomorphic imaging contrast (Fig. 1c). PG containers were either 3D printed or made of Barex^{TM}. By a reproducible shift and rotation of the inner structures, a patient's inter-fractional anatomy change was simulated and the applied dose to the tumor and OAR with and without an adaption of the treatment plan was evaluated in 3D. Results showed a very good agreement of PG evaluation with calculated treatment plans with passing rates >93% in 3D γ-analyses. The online plan adaption successfully compensated under-/over-dosages (down to 45%/ up to 180% of the prescribed dose) in the tumor/OAR that occurred without plan adaption.

ii. ADAM-PETer. Inter-fractional motion was simulated in a realistic setup using the ADAM-PETer phantom (Fig. 1e) [6]. To test a fractionated prostate irradiation scheme at a conventional linear accelerator (Artíste, Siemens Healthineers, Erlangen, Germany) with a microboost irradiation of an intraprostatic lesion (Fig. 1f), the phantom was equipped with a 3D printed and PG-filled prostate. The treatment fractions were irradiated directly after each other with the same PG container in the phantom. Between fractions, the position of the prostate was altered by changing the filling of an endorectal balloon. As expected, an under-dosage and a dose smearing within the microboost-volume were measured. Additionally, we could demonstrate significant inter-fractional shifts of the microboost volume of up to 5mm by the gel evaluation.

3.2.2. Intra-fractional motion. In case of intra-fractional patient motion (e.g. breathing), new IGRT treatment methods such as gating or tracking may be used for compensation. To test and validate an existing gating procedure at the MR-Linac, two motion phantoms were equipped with a PG-filled Barex^{TM} container simulating the tumor.

i. Geometric motion phantom. In a first experiment, the gating procedure at a clinical MR-Linac was simulated with a geometrical motion phantom consisting of a water-filled cylindrical case (d=22cm, l=50cm) holding a movable smaller water-filled cylinder (d=9cm, l=50cm) that can be loaded with the PG-filled Barex^{TM} tumor (Fig. 1g). The phantom motion was based on a $\cos^4$ trajectory with a peak-to-peak amplitude of 1.5cm. The gel was evaluated dosimetrically. Results for a gated treatment showed a homogenous target coverage similar to that of a static case with high 3D γ passing rates of >98.6%. Irradiation without motion compensation resulted in poor dose coverage with a γ_{3D} passing rate of 68.6%.

ii. Porcine lung phantom. To test the gated treatment at a clinical MR-Linac in a realistic setup, we used a porcine lung phantom (Fig. 1i) [2]. The PG tumor was sewed onto the mediastinum of the lung, which moved according to a real patient's breathing pattern. The target structure could be tracked by MRI throughout the entire treatment session. Additionally, homogeneous dose coverage without significant under or over-dosage was found indicating a successful gated irradiation. Dosimetric evaluation yielded a high 3D γ passing rate of 95.9%.

4. Discussion and Conclusion

In this paper, we present a variety of phantoms equipped with PG, which were used in clinical routine to perform both hardware and workflow tests in IGRT. This includes 3D isocenter alignment measurements at a clinical MR-Linac and end-to-end tests of online adaptive treatment procedures for both inter- and intra-fractional motion management. One important step to achieve a realistic setting is the use of 3D printed PG containers. This allows for adapting and extending existing phantoms to applications that require 3D dose measurements in irregularly shaped structures, e.g. for the simulation of specific organ sites. Furthermore, experiments have also shown that a geometric evaluation of the PG is possible directly after the irradiation. If, however, a full dosimetric evaluation is required, it is a common procedure in PG dosimetry to await the stabilization time and to normalize the required calibration curve at a reference point with a known dose determined e.g. by an ionization chamber (IC). However, due to the lack of bores in our anthropomorphic phantoms, independent absolute dose measurements with an IC were not possible. Hence, we based the reference dose on the treatment plan. To overcome this limitation, a recently developed technique combining thermoluminescence detectors (TLD) with PG can be used. Due to their compact design, TLDs can be attached on the surface of various phantom structures (OARs and/or tumor) serving as a reference system for absolute dosimetry [7,8]. Overall, the presented techniques and phantoms provide a flexible application of PG for 3D dosimetry in complex radiotherapy treatment techniques.

5. Acknowledgements

This work was supported by the EMPIR programme co-financed by the Participating States and from the European Union's Horizon 2020 research and innovation programme. We would like to thank the German Research Council (DFG) for the funding of the MR-Linac device.

6. References

[1] Baldock C et al. 2010 *Phys. Med. Biol.* **55** R1-63
[2] Mann P et al. 2017 *Phys. Med. Biol.* **62** 573–95
[3] Elter A et al. 2019 *Phys. Med. Biol.* **64** 04NT02
[4] Dorsch S et al. 2019 *Phys. Med. Biol.* **64** 205011
[5] Elter A et al. 2019 *Phys. Med. Biol.* **64** 225003
[6] Gillmann C et al. 2021 *Med. Phys.* **48** 1624–32
[7] Mann P et al. 2019 *Phys. Med. Biol.* **64** 45010
[8] Schwahofer A et al. 2020 *Phys. Med. Biol.* **65** 215002

Development of a novel proton CT system using a 3D scintillator detector

S Beddar, C D Darne, and D G Robertson
Department of Radiation Physics, The University of Texas MD Anderson Cancer Center, Houston, TX 77030, USA
Department of Radiation Oncology, Mayo Clinic, Phoenix, AZ 85259, USA

Email: abeddar@mdanderson.org

Abstract The goal of this work is to develop a novel proton computed tomography (CT) system using a 3-dimensional (3D) scintillator detector. The detector design of the pCT system consists of a 3D scintillator that generates scintillation light in response to the incident proton beams. A scientific charge-coupled device (CCD) camera placed along the beam axis captures a 2D projection or radiograph of the resulting 3D scintillation light signal. A set of CCD cameras placed laterally to the beam axis capture two unique depth dose distributions of the beam. These additional beam images can be used to improve the resolution of the proton radiograph. In this study, we demonstrate the feasibility of this unique imaging system design for proton radiography using a 3D liquid scintillator detector.

1. Introduction

The therapeutic potential of proton radiotherapy has not been fully realized because of limitations in precisely determining proton ranges within patients. Treatment planning systems for proton therapy rely on x-ray computed tomography (CT) to estimate proton ranges. However, this approach leads to uncertainties in proton range calculations and can cause excessive radiation dose to healthy tissues [1]. A proton-based imaging system would eliminate these range uncertainties and improve tumor targeting. Proton imaging systems generate images by measuring energy losses suffered by protons within the tissue. The proton stopping power ratios (SPRs) are then calculated by measuring the water equivalent thickness (WET) of the tissue. Proton integrating imaging systems determine the proton beam range by measuring the exit fluence using a planar or thin (2D) detector. However, in order for the planar detector to capture the Bragg peak, the proton beam range needs to be modulated. A 3D scintillator volume, on the other hand, has the ability to capture complete Bragg curves over a range of beam energies, which can make range modulation of proton beams redundant. This quality of the 3D detector is utilized in this study to design a novel proton CT (pCT) system with the potential to reduce the total dose and time required for proton imaging.

2. Materials and methods

The pCT system consists of an object/phantom to be imaged placed between the proton source and a monolithic solid scintillator (figure 1). The scintillation light produced within the scintillator in response to the proton beams is captured by 3 digital cameras from 3 mutually perpendicular directions. Camera 3, placed along the beam's-eye-view direction, generates the proton radiograph by integrating the light fluence. Cameras 1 and 2 generate lateral beam images that capture the beam location and its residual range within the scintillator.

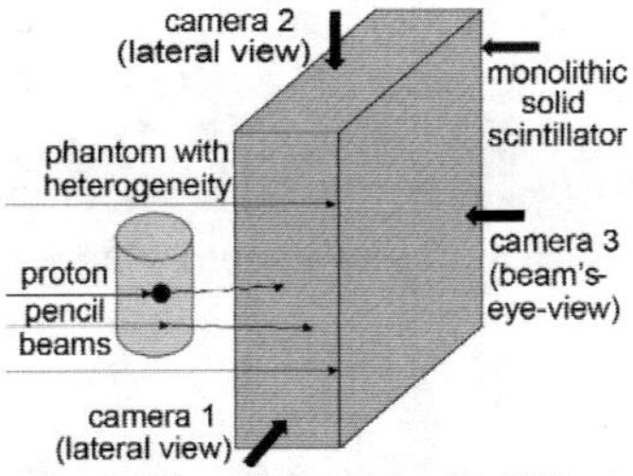

Figure 1. A schematic of the novel pCT system design.

For testing the design concept, a proton radiography system consisting of a 20 × 20 × 20 cm^3 tank filled with an organic liquid scintillator was built (figure 2a) [3]. The scintillation light from within the tank is captured by a charge-coupled device (CCD) using a 45° angled mirror. The system concept can be understood by considering a water-filled cube (50 mm^3) placed in front of the tank and exposing it to a passive scattering beam (figure 2a). The result will be a pull-back of the Bragg peak (BP) over the area of the cube, while the beam outside of the cube area remains unaffected. Figure 2b shows a representative Bragg curve captured by laterally placed cameras 1 and 2. The BP of the beam travelling through the cube will be pulled back, as represented by removing the red-shaded region from the depth-intensity curve. Along with changes to the BP location, the integrated light signal captured by camera 3 is also reduced, as revealed by plotting the cumulative light profile as a function of depth in figure 2c. Thus, the cumulative light intensity of the beam that traveled through the cube (indicated by the red dot) is lower than that of the beam that did not travel through it. This reduced cumulative light intensity corresponds to the WET of the cube.

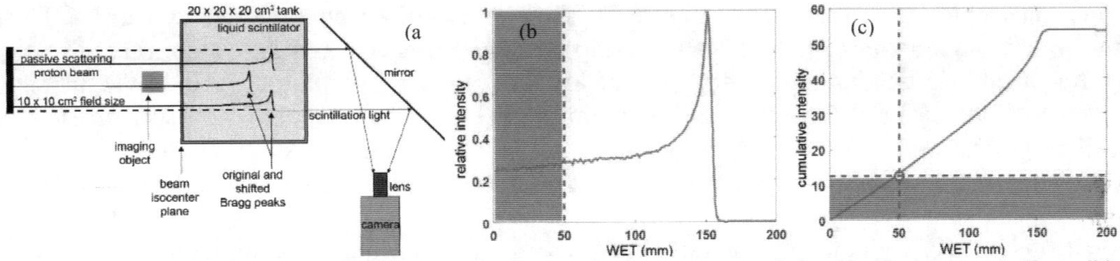

Figure 2. (a) A schematic of the proton radiography system used for preliminary studies. (b) A normalized proton beam Bragg curve is pulled back by a depth equivalent to the cube's water equivalent thickness. (c) The cumulative intensity curve measured by camera 3.

A Monte Carlo simulation of the system was performed to demonstrate the utility of using a 3D scintillator and multiple cameras for proton radiography. The passive scattering proton beam delivery system at The Univ. of Texas MD Anderson Cancer Center was modeled using the TOPAS MC system. The double scattering system, the beam collimation system, and other components of the treatment nozzle were modeled. The standard electromagnetic physics package (option 4) and the hadronic physics package QGSP_BIC_HP were selected. The simulation study included a passively scattered 148.8-MeV proton beam passing through an acrylic cylindrical phantom (1.18 g/cm^3, 5-cm diameter and height) placed in front of a 20-cm^3 water tank that mimicked the 3D scintillator.

Either a passive scattering or a scanned pencil beam can be used for radiography. For this study, proton imaging was conducted at the Proton Therapy Center at The University of Texas MD Anderson Cancer Center (PTCH) using a passive scattering beam with a 10 x 10 cm^2 field size. The proton images were generated with a 180 MeV energy. Optical artifacts including background signal and hot pixels were corrected for in each captured image [2]. In order to be useful for treatment planning purposes, the system must calculate tissue WETs from the measured signal intensities with high accuracy. To achieve this, a calibration curve was generated by stacking multiple plastic water slabs of known thicknesses in front of the 3D scintillator. The system testing was conducted using 2 phantom types – (a) a 3D-printed miniature skull phantom for qualitative imaging, and (b) 3 cylindrical inserts (solid water, acrylic, and cortical bone) from a CT Hounsfield Unit phantom (Sun Nuclear Corporation, Melbourne, FL) to check the system's accuracy for retrieving the phantom WETs. An initial proton CT image was obtained by placing the skull phantom on a rotating stage and acquiring proton radiographs using 66 projection angles. The CT was reconstructed using a filtered backprojection approach incorporating a Hamming filter.

3. Results/Discussion

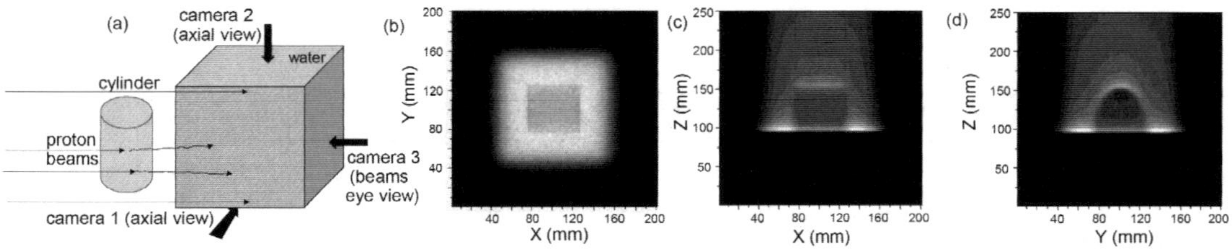

Figure 3. A simulation study of the novel proton CT system design. (a) The setup used for simulating the system. (b) A proton radiograph captured by camera 3. (c,d) Images of the beam dose distributions captured by cameras 1 and 2, respectively.

Figure 3 shows a simulation of the dose distributions captured by the cameras. The water tank mimicking the 3D scintillator detector produces a dose distribution that reflects the attenuation and scattering of the proton beam after it has traveled through the phantom. All 3 cameras capture unique projections of the dose deposited within the scintillator. Figure 3b shows the radiograph captured by camera 3 while figures 3c and 3d show the dose distributions captured by cameras 1 and 2, respectively. The signal integrating approach to radiography using a single camera (camera 3) captures a bulk of the light signal produced within the scintillator. This signal contains a mixture of proton paths and their corresponding residual ranges, and can subsequently result in loss of proton path information required to correct for proton scatter within the phantom. The use of 2 lateral cameras offer additional information about the scanned pencil beams including their 3D locations, divergence, and residual ranges which can be used for correcting proton scatter and thus improving the resolution of acquired radiographs.

Figure 4a shows the orientation of the 3 stacked Gammex inserts with respect to the beam direction. The theoretical WET values for the Gammex inserts were calculated based on their respective material densities, SPRs, and physical dimensions. A 2nd order polynomial fit to the measured calibration curve was performed and was used for calculating WETs (unit: mm) for each pixel of the recorded radiograph. Figure 4b shows the system acquired WET image of the stacked phantoms. The relative percentage accuracies in calculating WETs for solid water, acrylic, and cortical bone were −0.82%, 0.91%, and −2.43%, respectively [4]. Figure 4c shows a plot of WET line profiles across the inserts. The smooth edges of the profiles are likely due to the cone beam projection of the inserts resulting from a combination of proton beam divergence and camera system divergence [5]. A spatial resolution of 0.2 line-pair/mm (lp/mm) was determined for this system by imaging a phantom with varying line-pair densities [4].

Figure 5a shows a picture of the proton radiography setup on the patient couch at the PTCH. A qualitative observation of the skull phantom radiograph (figure 5b) demonstrates the ability of the system to image complex features without having applied any proton scatter corrections. Figure 5c shows a slice of the reconstructed skull phantom.

4. Conclusion

In this work, we report the concept, design, and development of a novel imaging system that uses a 3D scintillator detector and a beam-integrating approach to pCT. We presented the first ever proton radiographs and proton CT images using the beam-integrating 3D scintillator detector at a clinically relevant proton energy (180 MeV). The proton radiographs were generated in a single proton beam exposure and with an imaging time of 300 ms. The imaging system demonstrated an initial image resolution of 0.2 lp/mm, which can be further improved by using lateral camera views after implementing proton scatter corrections. While preliminary studies with homogeneous phantoms determined a good WET accuracy (< 2.5%), further investigation with complex and clinically relevant phantoms need to be conducted. Unlike most proton imagers, the 3D scintillator-based imager has a relatively simple design and is compatible with clinically available proton beam currents. It therefore has the potential to be quickly adapted into the proton therapy clinics to improve proton treatment planning for patients.

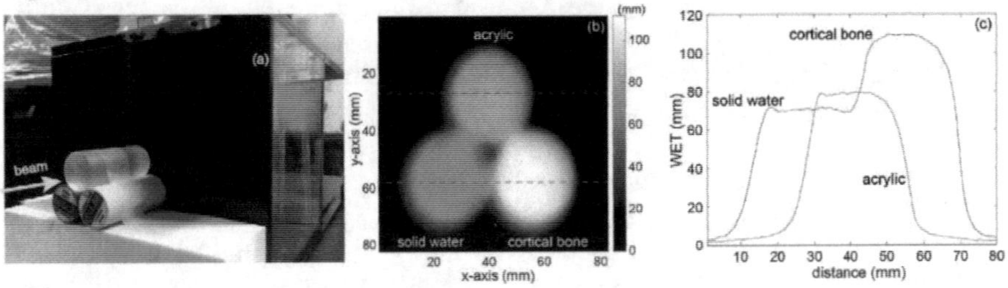

Figure 4. (a) A photograph showing the orientation of the Gammex inserts (solid water, acrylic, and cortical bone) used for radiography. (b) A proton radiograph of the inserts, with their WETs in mm. (c) WET profiles across the inserts as shown by the dotted lines in (b) [4].

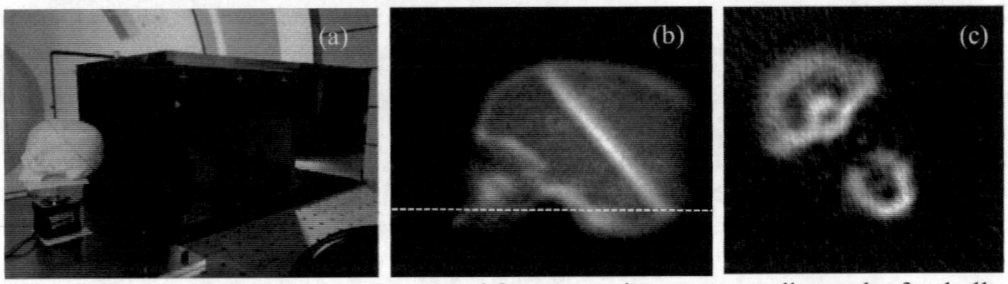

Figure 5. (a) A picture of the imaging setup used for generating proton radiograph of a skull phantom. (b) The proton radiograph of the skull phantom generated by the imaging system. (c) A slice of the reconstructed skull phantom along the dotted line highlighted in (b).

5. Acknowledgements

The authors thank Narayan Sahoo, PhD, for his help with measurements. The research reported in this publication was supported by the National Cancer Institute/National Institutes of Health under award number R01CA182450 and an Institutional Review Grant from MD Anderson Cancer Center. The authors have no relevant conflicts of interest to disclose.

6. References

[1] Moyers MF, Sardesai M, Sun S *et al* 2010 *Med. Dos.* **35** 179-94
[2] Robertson D, Hui C, Archambault L *et al* 2014 *Phys. Med. Biol.* **59** 23-42
[3] Darne CD, Alsanea F, Robertson DG *et al* 2017 *et al Phys. Med. Bio.l* **62** 5652-67
[4] Darne CD, Alsanea F, Robertson DG *et al* 2019 *Biomed. Phys. Eng. Express* **5** 045032
[5] Hui C K, Robertson D, Alsanea F *et al* 2015 *et al Biomed Phys Eng Express* **1** 025204

End-to-End Quality Assurance of Stereotactic Radiation Therapy Using an Anthropomorphic Head Phantom

KM Alexander[1], KH Dekker[1], T Olding[1,2,3], LJ Schreiner[2,3]
[1] Cancer Centre of Southeastern Ontario at Kingston Health Sciences Centre, Kingston, ON, K7L 5P9, Canada
[2] Department of Physics, Queen's University, Kingston, ON, K7L 3N6, Canada
[3] Department of Oncology, Queen's University, Kingston, ON, K7L 3N6, Canada

Email: kevin.alexander@kingstonhsc.ca

Abstract. In the clinic, routine quality assurance tests ensure the proper functioning of each individual aspect of a radiation therapy treatment delivery. However, these tests may not guarantee an accurate treatment delivery. In this work, we present the design and application of a head phantom (using 3D printing technology) developed for end-to-end quality assurance of a stereotactic radiation therapy treatment for brain metastases, coupled with inserts for ion chamber, and film and gel dosimeters. The head phantom was subjected to the entire clinical workflow, with each stage of the process being performed by the appropriate clinical personnel to ensure that this quality assurance test mimics the clinical scenario.

1. Introduction

Approximately 20-40% of all patients with cancer will develop brain metastases at some point in their life [1], with many of these patients suffering from multiple brain lesions [2]. One option for the treatment of this type of cancer is fractionated stereotactic radiation therapy (FSRT) using a linear accelerator (linac) [3]. Due to the high complexity of these deliveries and the small targets involved, it is crucial to ensure the geometric and dosimetric accuracy of the delivery. While routine quality assurance (QA) tests are performed to verify that treatment machines and planning systems are working properly, this does not inherently mean that the overall treatment process and delivery is accurate. For this reason, initiating routine end-to-end (E2E) QA would be useful to validate the entire treatment process [4,5].

In this work, we performed an in-house E2E QA test of a FSRT brain metastases volumetric modulated arc therapy (VMAT) radiation therapy treatment by creating an anthropomorphic head phantom, handing it over to the radiation therapists, and verifying the entire process (from CT simulation, through patient setup, to radiation delivery) by measuring the overall dosimetric and geometric accuracy using multiple dosimeters.

2. Methods

2.1. Phantom Design

An anthropomorphic head phantom was created and adapted so that it could accommodate a variety of dosimeters, while also approximating the densities and CT numbers of a human head. To create the phantom, a 3D model of the Rando phantom skull was first generated. To accommodate dosimeters, a cylindrical cavity (diameter = 7.0 cm, length = 12.6 cm) was added to the skull, running in-line with the neck (Figure 1). The skull model was post-processed to ensure that the brain region was watertight in order to replicate the radiological properties of a human brain. The skull model was 3D printed using a high density iron-filled PLA (3D Printing Canada, Hamilton, Ontario) using a Prusa i3 MK3 3D printer (Prusa Research, Prague, Czech Republic) at 100% infill. The skull was filled with water and sealed. Polydoh moldable plastic (Materialix, Jaslo, Poland) was used to create some portions of the phantom simulating soft tissue.

2.2. Dosimetry Systems and Phantom Inserts

A cylindrical insert (diameter = 7.0 cm, length = 14.0 cm) was 3D printed (100% infill) using PLA (Prusa Research) with a 10 cm deep channel in the centre for an ion chamber. A Capintec PR-05P 0.07 cc thimble ionization chamber (Capintec, Ramsey, NJ) was used in this work. The chamber insert was placed in the head phantom, and the ion chamber was positioned in the centre of the insert.

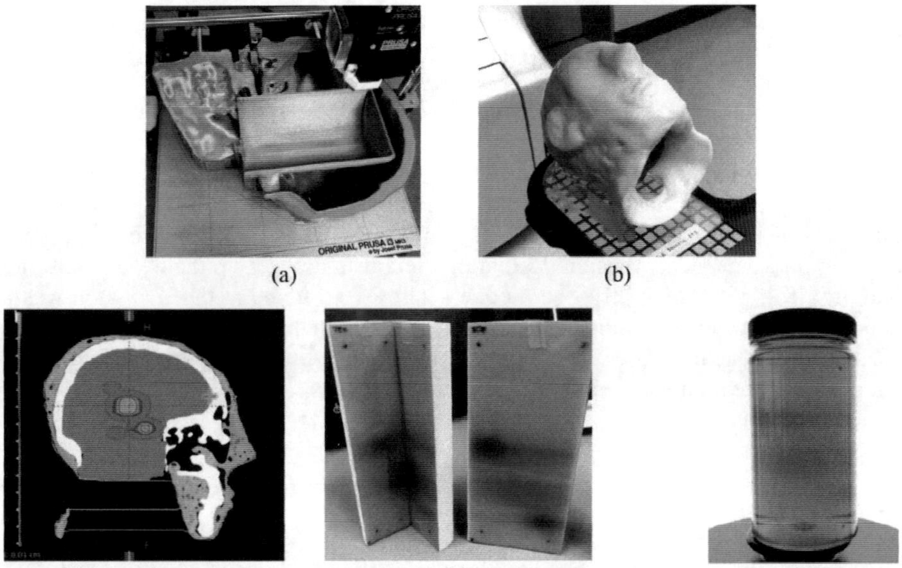

(a) (b) (c) (d) (e)

Figure 1. (a) The 3D printed skull at the midway point of the print, showing the cylindrical cavity and hollow brain. **(b)** The head phantom, placed on a neck rest on the linac couch. **(c)** Eclipse treatment planning system screen capture showing the brain region (no cavity/insert visible) to highlight the contoured regions. **(d)** The film insert, showing the three irradiated film plane PTVs and the fiducial points on each film. **(e)** An FXG gel dosimeter after irradiation, showing the irradiated dose distribution and optical fiducial markings.

A cylindrical insert for film (Gafchromic EBT3, International Specialty Products, Wayne, NJ) was 3D printed (PLA, 100 % infill) in four parts to allow for three films to be placed in three separate planes. Metal BBs were implanted in the insert sections at the surface of each 3D printed film plane and marker dots coincident with the BBs were marked on the film to facilitate registration. Film readout was performed using our in-house point scanner and scans were acquired 24 hours after irradiation at 0.6 mm resolution. Analysis was performed using 3D Slicer and the Film Dosimetry Analysis slicelet [6].

The third phantom insert was a gel dosimeter jar. Both leuco crystal violet (LCV) micelle gel dosimeters [7] and Fricke-xylenol orange-gelatin (FXG) gel dosimeters [8] were used in this work. All gel dosimeters were scanned using a Vista Optical CT Scanner (ModusQA, London, ON) pre-irradiation, and 30 mins post-irradiation. Analysis of the gel dosimeter data was performed using 3D Slicer and the Gel Dosimetry Analysis slicelet [9].

2.3. CT Simulation, Treatment Planning, and Delivery
To best replicate the clinical scenario, the head phantom was handed over to the radiation therapy team and was CT scanned according to our clinic's brain VMAT FSRT policy. The phantom was placed in a neck rest and a thermoplastic mask was used for immobilization on a type-S overlay head and neck board (Civco Medical Solutions, Coralville, IA). Separate CT scans were acquired for each of the dosimeter inserts using a Philips Brilliance Big Bore CT scanner (Philips Medical Systems, Cleveland, OH) with a tube energy of 120 kV and a slice thickness of 1 mm.

Treatment planning was performed by the Physics team using Eclipse (Varian Medical Systems, Palo Alto, CA). Four spherical brain metastases were defined: one 2 cm diameter metastasis and three 1 cm diameter metastases. Planning target volumes (PTVs) were generated following our clinical policy. Two coplanar 6 MV clockwise/counterclockwise arcs were used, followed by a third 6 MV arc, involving a couch rotation of 45°. Planning was performed so that 100% of the prescription dose covered at least 95% of each of the 4 PTVs. An EVAL50 structure was also generated, consisting of all four PTVs plus a 0.8 cm margin in the superior/inferior direction and 1.5 cm everywhere else. As per clinical protocol, the calculated 50% dose cloud was contained within this structure.

To fully assess the image-guided radiation therapy (IGRT) treatment process for a VMAT FSRT treatment, our clinical radiation therapy team performed the setup and delivery of the treatment. Each type of dosimeter was irradiated on a different day, meaning that there were three independent setups and radiation therapist teams who assisted. Radiation therapists first set-up the head phantom according to the clinical protocol. Next, they acquired a cone beam CT scan and a 3D-3D match was performed within a region of interest encompassing the skull. The translational and rotational shifts were calculated and applied as needed to achieve a clinically acceptable match. Once the phantom had been shifted to the treatment position, the plan was delivered for each dosimeter.

3. Results

3.1. Phantom Design and Set-Up
The head phantom developed in this work has three distinct anatomical features: brain, soft tissue, and skull. In the phantom, the mean CT number of the water-filled brain is 10 HU, the PCL skin and soft tissue layer (while slightly variable due to pockets of air trapped between layers of PCL) is 65 HU, and the 3D printed skull has a mean CT number of 750 HU. The gel dosimeter has a mean CT number of 10 HU, while the 3D printed PLA inserts have a mean CT number of 120 HU.

CBCT imaging was performed before each delivery. The applied translational shifts were < 2 mm and the rotational shifts were $< 0.5^{\circ}$, indicating that the initial set-ups by the radiation therapists were accurate using the setup markings on the mask.

3.2. Dosimeter to Plan Dose Analysis
As shown in Table 1, ion chamber dose measurements were within 2% of the mean dose calculated in Eclipse within the sensitive volume of the chamber. Across all film and gel fiducial based registrations, the average root mean square error (RMSE) between coincident fiducial points was < 1 mm. Film and gel dosimeters were all calibrated under standard conditions [9] so that an absolute dosimetric comparison could be made (no dose scaling was performed). Gel dosimeters were handled according to the protocols described in Olding and Schreiner (2011) [10]. In Figure 2, dose profiles across the center of each of the 4 PTVs are shown for film, FXG gel, and LCV gel. There was very good spatial agreement between the Eclipse planned dose and the calibrated film and gel measurements, however,

dosimetrically there are a few regions of disagreement in both film and gel when compared to Eclipse planned doses. Gamma comparisons for the film measurements show excellent agreement with the Eclipse planned dose at both 3%/3mm and 3%/1mm criteria (Table 2). While the calibrated film measurements agree with Eclipse planned doses for the 2 cm PTV, the 1 cm PTV peak doses are roughly 6% higher than the Eclipse dose. Dose readings in the calibrated gel measurements for both gel types are consistently low when compared to Eclipse and range from 3-6% low in peak dose ratios (Table 1). This dosimetric difference is also shown in the dose profiles in Figure 2.

Table 1. Dosimeter measurements vs. Eclipse. Mean dose ratio is the measurement relative to the calculated dose within the PTV.

Dosimeter	PTV # (Diameter)	Prescription Dose (Gy)	Mean PTV Dose Ratio (%)
Ion Chamber	1 (2 cm)	6.5	98.5 ± 0.5
Film	1 (2 cm)	2.5	102 ± 2
	2 (1 cm)		106 ± 2
	3 (1 cm)		106 ± 2
	4 (1 cm)		105 ± 2
FXG Gel	1 (2 cm)	2.5	96 ± 3
	2 (1 cm)		94 ± 3
	3 (1 cm)		94 ± 3
	4 (1 cm)		94 ± 3
LCV Micelle Gel	1 (2 cm)	13	97 ± 3
	2 (1 cm)		97 ± 3
	3 (1 cm)		96 ± 3
	4 (1 cm)		96 ± 3

Table 2. Gamma pass rates for film and gel dosimeters performed using two different sets of comparison criteria: 3%/3mm and 3%/1mm.

Dosimeter	Comparison Region (PTV # in film plane)	Gamma Pass Rate (3%/3mm)	Gamma Pass Rate (3%/1mm)
Film	Plane A (1)	100 %	98.9 %
	Plane B (2 & 4)	100 %	98.5 %
	Plane C (3)	100 %	93.3 %
FXG Gel	EVAL50	99.4 %	93.6 %
	Entire jar	99.3 %	94.0 %
LCV Micelle Gel	EVAL50	97.1%	50.2 %
	Entire jar	94.6%	55.6 %

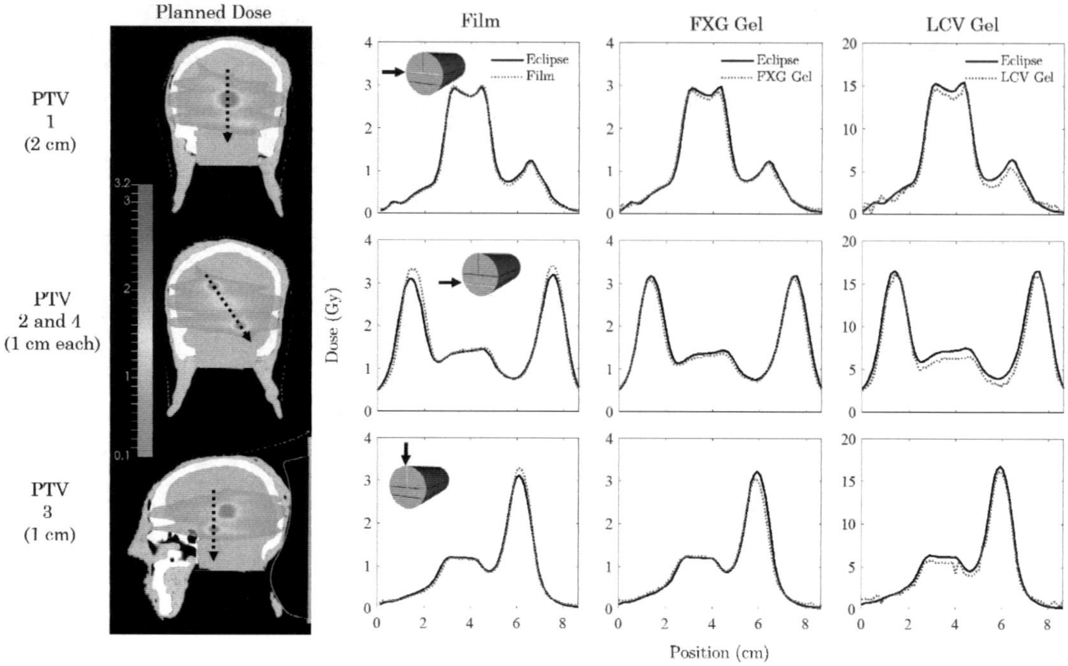

Figure 2. Dose profile comparisons along the dotted lines (as shown overlaid on the dose distribution) for irradiations to film, FXG gel, and LCV micelle gel. Film planes are indicated in the inset film insert diagram. Color bar ranges from 0.1 Gy to 3.2 Gy, representative of the deliveries to film and FXG gel, whereas LCV micelle gel has a maximum dose of ~17 Gy due to the different plan normalization (note the difference in the dose axes). Irradiations to each dosimeter were performed on separate days.

4. Discussion and Conclusions

The head phantom developed in this work for E2E QA closely resembles a human head, was easy for clinical staff to use at each stage of the treatment process and produced a realistic E2E QA test. The head phantom was economical to manufacture, with a materials cost of ~ \$300. Using the phantom, our clinic's radiation therapists were able to mimic the entire radiation therapy treatment process from patient CT simulation imaging, phantom set-up, CBCT imaging and alignment, to treatment delivery.

This multi-dosimeter approach allowed for verification of repeat treatment deliveries with validation in multiple dimensions. Ion chamber measurements showed excellent agreement with the planned dose delivery. Film dosimetry analysis using our in-house point scanner showed very good spatial agreement with the planned dose delivery. However, since our clinical policy dictates that the dose calculation be performed using a 2 mm dose calculation grid size, an underreporting of dose is observed when compared to our high resolution (0.6 mm) film measurements. A re-calculation using a 1 mm grid size was performed in Eclipse and the dose for each of the 1 cm PTVs increased by up to 5% (insignificant change to the peak dose to the 2 cm PTV), placing the film and Eclipse doses in excellent agreement. Gel dosimetry was geometrically accurate; however, it underreported the dose delivery due to optical light scatter issues within the optical CT scanner [11]. This discrepancy in dose is not as evident in the gamma pass rates shown in Table 2 due to the noisy nature of the optical CT gel measurements, leading to higher gamma pass rates [12].

The head phantom had some limitations, in that it lacked some bony anatomy (no spine due to the insert cavity), was unable to measure dose in the superior part of the brain using gel, and could not simulate patient motion during treatment. However, given these limitations and challenges using each dosimeter, together, all four dosimeters helped to give a full evaluation of the dose delivery.

Overall, these E2E tests helped to mimic a real patient IGRT process by employing specific clinical personnel to perform each step in the process, used multiple dosimetry systems, and ultimately provided a more complete picture of how the entire VMAT FSRT radiation therapy process is working properly to accurately deliver the planned treatment. Our results give us confidence that our clinical workflow is safe, accurate, and effective at treating small targets and sets the stage for routine E2E QA testing.

5. References

[1] Wiggenraad R, Kanter AVD, Kal HB *et al* 2011 *Radiother. Oncol.* **98** 292-7
[2] Lohr F, Pirzkall A, Hof H *et al* 2001 *Semin. Surg. Oncol.* **20** 50-6
[3] Nichol A, Ma R, Hsu F *et al* 2016. *Int. J. Radiat. Oncol. Biol. Phys.* **94** 312-21
[4] Schreiner LJ 2019. *J. Phys. Conf. Ser.* **1305** 012062
[5] Baldock C *et al* 2010 *Phys. Med. Biol.* **5** R1-R63
[6] Alexander KM, Pinter C, Fichtinger G *et al* 2017 *J. Phys. Conf. Ser.* **847** 012061
[7] Babic S, Battista J, Jordan K 2009 *Phys. Med. Biol.* **54** 6791-808
[8] Babic S, Battista J, Jordan K 2008 *Phys. Med. Biol.* **53** 1637-50
[9] Alexander KM, Pinter C, Fichtinger G *et al* 2018. *Biomed. Phys. Eng. Express* **4** 045041
[10] Olding T, Schreiner LJ 2011 *Phys. Med. Biol.* **56** 1259-79
[11] Dekker KH, Hazarika R, Jordan KJ 2018 *Phys. Med. Biol.* **63** 21NT02
[12] Schreiner LJ, Holmes O, Salomons G 2013 *J. Phys. Conf. Ser.* **444** 012016

Synthesis and structural characterization of Al_2O_3 nanoparticles: Towards 3D optically stimulated luminescence dosimetry

Camilla L Nielsen[1], Martin Bondesgaard[2], Rosana M Turtos[1], Brian Julsgaard[1,3], Bo B Iversen[2,3], Ludvig P Muren[4,5] and Peter Balling[1,3]
[1]Department of Physics and Astronomy, Aarhus University, Denmark
[2]Department of Chemistry, Aarhus University, Denmark
[3]Interdisciplinary Nanoscience Center (iNANO), Aarhus University, Denmark
[4]Danish Center for Proton Therapy, Aarhus University Hospital, Denmark
[5]Department of Medical Physics, Aarhus University & Aarhus University Hospital, Denmark

E-mail: cln@phys.au.dk

Advanced radiotherapy techniques, which plan and deliver a treatment in complicated 3D geometries with steep dose gradients, push 3D dosimetry with correspondingly high spatial resolution to the top of scientific and clinical agendas. This paper presents the first steps taken towards an inexpensive and reusable material for 3D dosimetry based on optically stimulated luminescence (OSL). Carbon-doped alumina (Al_2O_3:C) nanoparticles were synthesized using supercritical flow synthesis, in which product properties can be finely controlled. The particles were characterized using electron microscopy and powder x-ray diffraction. C-doping did not alter the crystallographic structure appreciably, and a high elemental signal from C could be measured. Nanoparticles of amorphous γ-Al_2O_3:C were achieved, however calcining these to produce the OSL-relevant α-phase yielded microparticles. Future work will aim at producing phase-pure α-Al_2O_3:C nanoparticles with a narrow size distribution below 10 nm, and controllable C-concentration and O-deficiency.

1. Motivation & Introduction

To spare healthy tissue near a cancer tumor, modern radiotherapy techniques, such as intensity modulated- , proton-, and heavy-ion radiotherapy, plan and deliver a treatment in a complicated 3D geometry with steep dose gradients. In order to assure the quality of the treatment, patient-specific or at least treatment-plan-specific dosimetry also in 3D with high spatial resolution should be performed. Historically, gel- and polymer matrices (e.g. radiochromic [1, 2] or polymerizing) have been used, however problems with their reusability and reproducibility have spurred renewed interest in dosimeters based on optically stimulated luminescence (OSL) that rely on physical properties rather than chemical reactions. This paper presents progress and considerations regarding an inexpensive and flexible dosimetry system based on OSL nanoparticles embedded in a transparent elastomer matrix, which has previously been suggested by our group [3, 4]. Such a system is reusable as it can be reset by emptying the populated trap states responsible for information storage (so-called bleaching) with either light or heat. Readout of this dosimeter can be performed by a movable laser sheet, which illuminates and stimulates a thin slice of the dosimeter at a time. By imaging this 2D slice onto a suitable sensor such as a CCD camera and then moving the laser sheet down through the dosimeter, the entire 3D shape can be read out (see [5] for a sketch of the readout setup).

Carbon-doped alumina in the crystallographic α-phase (α-Al_2O_3:C) is well-known for exhibiting highly sensitive OSL signals in a wide dose range [6]. To achieve high spatial resolution and transparency of a 3D OSL dosimeter, nanoparticles rather than bulk are preferable, in particular for high–refractive index materials like Al_2O_3 [7]. This article presents synthesis and characterization of Al_2O_3 nanoparticles in various phases with and without C-doping. To allow fine control of many properties, industry-inspired supercritical flow synthesis was used. Particles are investigated with elemental analysis, electron microscopy, and x-ray diffraction.

Batch-based techniques for synthesis of nanoparticles such as sol-gel or traditional hydrothermal synthesis are limited by relatively long reaction times, where both nucleation and growth occur at the same time. This results in wide size distributions or large particles (via the LaMer model [8]). To accurately control the size of small nanoparticles, a different approach is necessary. In this work, supercritical flow synthesis was used to produce nanoparticles of aluminum oxide. The supercritical state of matter can be achieved by heating while at the same time applying pressure to a liquid. The result is a fluid with properties (solubility, diffusivity, density) between those of a liquid and gas. In supercritical flow synthesis, the decreased solubility of precursor molecules in supercritical solvent compared to the liquid state leads to flash nucleation of nanoparticles [9]. A sketch of the flow reactor can be seen Figure 1. The primary reward of such a setup is full control of both the precursor concentration, flow rates of both precursor solution and solvent as well as the pressure and temperature at the T-mixing section and in the maturing reactor. In this way, the size distribution, shape and phase of the nanoparticles can be largely controlled [9].

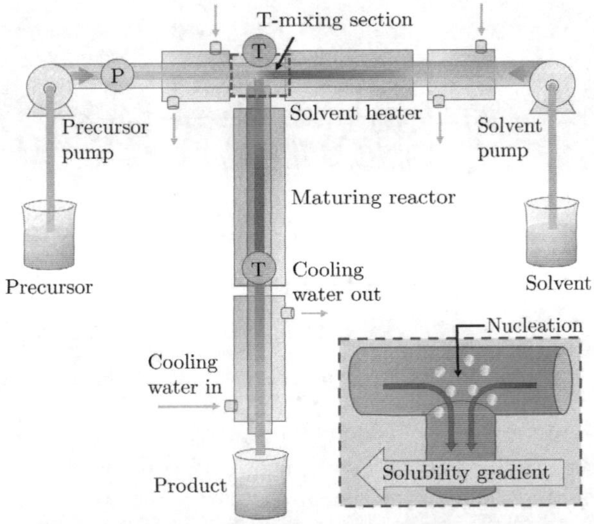

Figure 1. Sketch of the supercritical flow synthesis setup used in this work. The setup consists of pressurized tube lines for the precursor and solvent on each side, as well as a central section, where they are mixed and the reaction takes place. In the lower right corner is a close-up of the T-mixing section, where nucleation takes place. The just-seeded nanoparticles are swept into a maturing reactor, where growth can be precisely controlled by adjusting temperature and flow rates. The particles are cooled before being collected. A barometer or thermocouple is indicated by a P and T, respectively. Adapted from [9].

Journal of Physics: Conference Series **2167** (2022) 012023 doi:10.1088/1742-6596/2167/1/012023

2. Experimental Details

Synthesis of both pure and C-doped Al_2O_3 nanoparticles was performed using the setup seen in 1 with a precursor solution of $Al(NO_3)_3$. Samples from pure precursor and with an added C-source, corresponding to an expected doping of 4 atomic-%, were prepared for direct comparison between undoped and C-doped particles. At the synthesis temperatures relevant for this work, two competing reactions exist, namely the synthesis of boehmite, γ-AlO(OH), or alumina, γ-Al_2O_3. Both these phases can be calcined (heated) at temperatures above 1000 °C to produce corundum, α-Al_2O_3. To visualize the samples, scanning transmission electron microscopy (STEM) was performed (FEI Talos F200X), along with nanoscale elemental analysis by energy dispersive x-ray spectroscopy (EDS) measuring the characteristic x-rays produced by the electron beam. Structural characterization using powder x-ray diffraction (PXRD) was done on a Rigaku SmartLab II (Cu $K\alpha_1$-radiation, λ = 1.54056 Å).

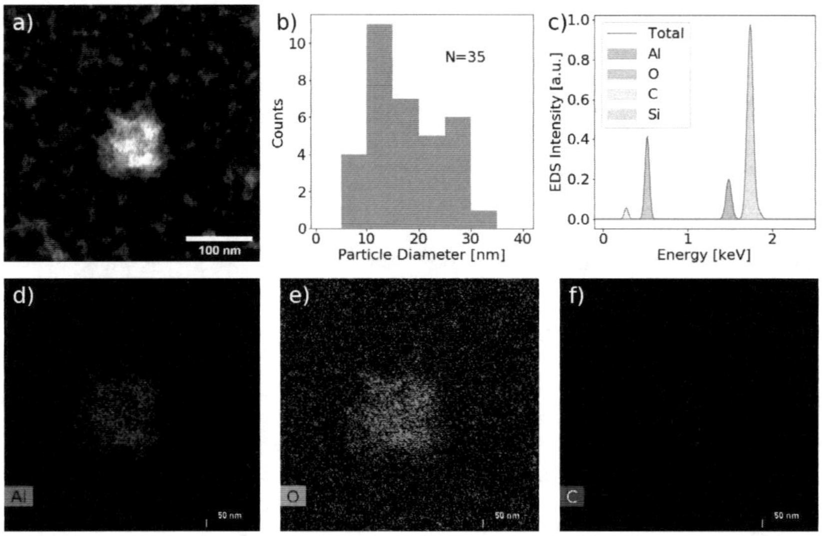

Figure 2. HAADF STEM (a) of C-doped Al_2O_3 particles. A size distribution (b) of particles measured visible in the HAADF-image and the EDS spectrum (c) is plotted. The large Si peak is from the TEM-grid. The color maps (d-f) show elemental EDS-intensities of Al, O, and C, respectively.

3. Results & Discussion

Results from the STEM-analysis of the C-doped particles can be seen in 2. An image taken in high-angle-annular-dark-field (HAADF) mode (2a) and EDS-maps (2d-f) for Al, O, and C are shown. The particles visible in 2a are clearly outlined in the EDS maps. Although the main feature is larger than the target of sub–10 nm, it seems to consist of a substructure of agglomerated nanoparticles, and around these are smaller particles, closer to the expected size as seen in the histogram of 35 particles visible in the HAADF-image (2b). The C-signal may not look high in the EDS-map or -spectrum (2c) due to a lower interaction cross section for light elements, but the concentration is in fact much higher than expected. Atomic composition of the particle is 20.03 $\pm$ 1.11% Al (36.00%), 57.99 $\pm$ 1.79% O (60.00%), and 21.97 $\pm$ 0.76% C (4.00%) (expected in parentheses, at-%). There is no detectable trace of N in the doped sample, but a small signal is seen in the undoped sample (7.52 $\pm$ 1.93 at-%), probably in the form of nitrates, suggesting that doping may facilitate full reaction of $Al(NO_3)_3$. The high C-concentration is likely due to organic residues from the solvent on the surface, in spite of extensive washing. To avoid this contaminant, supercritical water could be used in future synthesis. The high background O-signal is likely oxidation of the Si surface of the TEM grid.

PXRD reveals the structure of as-synthesized nanoparticles to be primarily amorphous both for the undoped and C-doped samples. This can be seen in 3a. The few discernible peaks are very broad, indicating nanosized crystallites in agreement with the STEM images. For reference, theoretical Bragg peaks of γ-Al_2O_3 and γ-AlO(OH) are plotted as tick marks, but poor data quality has made an actual refinement impossible. Although hard to say for certain, the particles seem to be a mixture of the two phases in both cases. This can explain the large particle seen in the STEM image (2a), because γ-AlO(OH) often acts as a binder between crystallites [10]. Most importantly, the diffractograms show that a C-doping of 4% does little to change the structure. The combined analysis of PXRD and STEM indicates small crystallites of γ-Al_2O_3 and γ-AlO(OH) merged together into large particles by amorphous γ-AlO(OH).

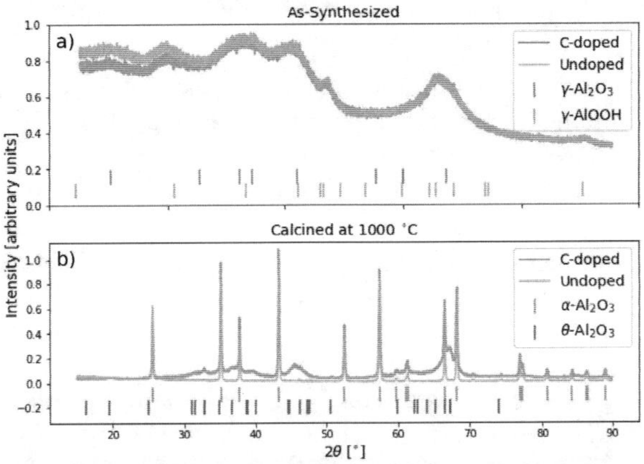

Figure 3. PXRD difractograms of C-doped and undoped Al_2O_3 particles as-synthesized (a) and after heating to 1000 °C (b). C-doped is plotted in red and undoped in blue in both cases. Also plotted for reference are tick marks for theoretical Bragg peaks of γ-Al_2O_3 [11] and γ-AlO(OH) [12] (top) as well as α-Al_2O_3 [13] and θ-Al_2O_3 [11](bottom).

After heating to 1000 °C, the particles become highly crystalline (3b). Reference Bragg positions plotted at the bottom (in this case α-Al_2O_3 and θ-Al_2O_3), reveals that the undoped sample is pure α-Al_2O_3, while the C-doped sample has a small contamination probably of θ-Al_2O_3. The width of the peaks has contracted dramatically from the broad features in the as-synthesized diffractograms. This signifies microparticles rather than nanoparticles. Together with evidence from the STEM analysis, it could be suggested that a microparticle, consisting before heat treatment of smaller crystallites merged together by amorphous material, grows into a single microcrystallite of primarily α-Al_2O_3 upon calcination. Further refinement and complementary STEM analyses of the calcined samples are necessary before a definite conclusion can be made.

Once developed, the composite 3D dosimeter consisting of these OSL active nanoparticles embedded in a suitable matrix material should in every voxel be comparable to a measurement done by a point dosimeter (such as an ion chamber). This entails a precision of ~ 2 % even at doses as low as 0.1 Gy measured in small voxels of the order of mm^3's [14]. Thus even sharp dose gradients in spatially complicated treatment plans will be within the detection limit. Furthermore, due to the nature of the energy storage mechanism involved, the dosimeter will be reusable and relatively insensitive to ambient conditions; as long as it is stored at room temperature in the dark, fading of the signal should be as low as 3 % per year [15]. Further work is warranted to study the effect of long-wavelength ambient work light, which would allow for convenient handling in a clinical environment. Similarly, various methods for bleaching and thereby resetting the dosimeter - such as thermal, UV, or a combination - should be studied.

As mentioned previously, the necessity of very small nanoparticles stems from the desire to minimize scattering of light through the dosimeter. Even though the refractive index of Al_2O_3:C is relatively high (~ 1.8) compared to that of most elastomers (e.g. silicone with 1.4), Mie scattering can be neglected if the particle size is less than 10 % of the wavelength of light. Remaining is Rayleigh scattering, which scales as [16]:

$$T = \exp\left(-\frac{32\pi^4 V_p \ell r^3 n_m^4}{\lambda^4} \left[\frac{(n_p/n_m)^2 - 1}{(n_p/n_m)^2 + 2}\right]^2 \right),$$

where T is the transmission at wavelength λ through a sample of length ℓ with a volume percentage V_p of nanoparticles with radius r. n_p is the refractive index of the nanoparticles, whereas n_m is that of the matrix material. Evidently the two most important factors in increasing the transmission is lowering the particle size, matching the refractive indexes and/or lowering the volume percentage. Naturally, there is a lower boundary for the volume percentage so that the signal level is still sufficient and the spatial resolution is not limited by the separation between particles.

4. Conclusion

In conclusion, ideal synthesis parameters to optimize the nanoparticle size distribution and avoid agglomeration have yet to be found. Also, washing procedures must be improved to avoid organic contamination of the surface, which can lead to uncontrollable C-doping during calcination. Furthermore, a different calcination procedure must be used, where crystal growth into microparticles can be avoided. The studies do reveal the possibility of synthesizing nanoparticles down to below 10 nm, especially evident in the C-doped sample, and calcination yielding phase-pure α-Al_2O_3. Apart from the C-concentration, the OSL response of α-Al_2O_3:C also depends on O-deficiency of the lattice resulting in F and F^+-centers, which are believed to be responsible for radiative recombination [17]. These have a high energy of formation [18], which is why conventional samples are produced at very high temperatures (> 2000 °C). However, as seen very clearly from calcinating the nanoparticles synthesized in the present work, such temperatures are - with current procedures - incompatible with very small particles. The challenge still remains to produce nanosized Al_2O_3:C with the correct level of O-deficiencies to achieve the desired OSL properties. It is unknown whether the α-phase of Al_2O_3:C is crucial for OSL to be present, or indeed if the directly synthesized γ-phase can also be utilized as long as the correct level of O-vacancies is present.

In summary, a number of challenges and investigations are still ahead before such a 3D OSL dosimeter is ready for clinical use - from the synthesis of ideal nanophosphors to the embedment of these at volume percentages suitable for both response level and transparency. However, the quest is strongly motivated by the potential for high-accuracy, reusable, and affordable 3D quality assurance of state-of-the-art treatment plans.

5. Acknowledgements

This research was supported by the Novo Nordisk Foundation. C. L. N. would like to thank Mohammad Aref Hasen Mamakhel for help with calcination and Sanna Sommer for discussions and assistance with PXRD.

6. References

[1] Høye EM, Balling P, Yates ES *et al* 2015 *Phys. Med. Biol.* **60** 5557-70
[2] Høye EM *et al* 2016 *J. Phys. Conf. Ser.* **755** 1-5
[3] Sadel M, Høye EM, Skyt PS *et al* 2017 *J. Phys. Conf. Ser.* **847** 012044

[4] Balling P, Høye EM, Sadel M *et al* 2016 *European Patent Office* 1-18 Available: http://scholar.google.com/scholar?hl=en{\&}btnG=Search{\&}q=intitle:Designated+Extension+States+:{\#}0
[5] Nyemann JS *et al. J. Phys. Conf. Ser.* In press.
[6] Akselrod MS, Kortov VS, Kravetsky DJ *et al* 1990 *Radiat. Prot. Dosim.* **32** 15-20
[7] Jin Y *et al* 2016 *J. Mater. Chem. C* **4** 3654-60
[8] LaMer VK, Dinegar RJ *et al* 1950 *J. Am. Chem. Soc.* 72 4847-54
[9] Hellstern HL, Becker J, Hald P *et al* 2015 *Ind. Eng. Chem. Res.* **54** 8500-8
[10] Pyda W 2016 *Ceramic Materials* **64** 376-83
[11] Zhou RS, Snyder RL 1991 *Acta Crystallographica Section B* **47** 617-30
[12] Milligan WO, Mcatee JL 1956 *Am. J. Phys. Chem.* **60** 273-77
[13] Finger LW, Hazen RM 1978 *J. Appl. Phys* **49** 5823-6
[14] Jensen ML, Turtos RM, Nyemann JS *et al. J. Phys. Conf. Ser.* In press.
[15] KortovV 2007 *Radiat. Meas.* **42** 576-81
[16] Jin Y *et al* 2016 *J. Mater. Chem. C* **4** 3654-60
[17] McKeever SWS, Akselrod MS *et al* 1999 *Radiat. Prot. Dosim.* **84** 163-6
[18] Dienes GJ, Welch DO, Fischer CR 1975 *Phys. Rev. B* **11** 3060-70

From conception to clinical trial: IViST, the first multi-sensor-based platform for real-time In Vivo Source Tracking in HDR brachytherapy

H. M. Linares Rosales[1,2], Audrey Cantin[1], Sylviane Aubin[1], Sam Beddar[4], Luc Beaulieu[1,2]

[1]Département de physique, de génie physique et d'optique et Centre de recherche sur le cancer, Université Laval, Québec, Canada.
[2]Département de radio-oncologie et Axe Oncologie du CRCHU de Québec, CHU de Québec - Université Laval, QC, Canada.
[3]Department of Radiation Physics, The University of Texas MD Anderson Cancer Center, Houston, TX, United States.
[4]The University of Texas MD Anderson UTHealth Graduate School of Biomedical Sciences, Houston, TX, United States.

Email: Luc.Beaulieu@PHY.ULAVAL.CA

Abstract. This study is aimed to introduce a novel multi-sensor-based dosimetry platform for real-time plan monitoring in HDR brachytherapy: IViST (In Vivo Source Tracking). IViST is a platform composed of three components: 1) an optimized and characterized multi-point plastic scintillator dosimeter (3 points mPSD; using BCF-60, BCF-12, and BCF-10 scintillators), 2) a compact assembly of photomultiplier tubes (PMTs) coupled to dichroic mirrors and filters for high-sensitivity scintillation light collection, and 3) a Python-based graphical user interface used for system management and signal processing. IViST can simultaneously measure dose, triangulate source position, and measure dwell time. By making 100 000 measurements/s, IViST samples enough data to quickly perform key QA/QC tasks such as identifying wrong individual dwell time, position, or interchanged transfer tubes. By using 3 co-linear sensors and planned information for an implant geometry, the platform can also triangulate source position in real-time. A clinical trial to validate this system is presently on-going using the IViST system.

1. Introduction

In vivo dosimetry (IVD) in brachytherapy (BT) aims to quantify in real-time the agreement between the treatment plan and delivered dose. Presently, BT clinics do not verify their treatments in real-time: (a) commercial real-time IVD systems have small signal-to-noise ratios (SNR) (limited time resolution, large measurements uncertainties, detection of only large errors); (b) laboratory real-time systems can be cumbersome to operate (more difficult to disseminate to BT clinics and workflow); (c) non PSD based detector systems do not employ sophisticated error detection algorithms, large false error rate, poor confidence in error reporting. Several studies [1–8] have been focused on developing detectors and methods for real-time source monitoring in BT. The advantages of using plastic scintillator detectors (PSDs) have been demonstrated and discussed in the literature [9, 10]. Their real-time response being an important advantage for this novel application. Nevertheless, PSDs are affected by stem effect and temperature variations [11, 12]. The proper optimization of the optical chain combined with the

Journal of Physics: Conference Series 2167 (2022) 012024 doi:10.1088/1742-6596/2167/1/012024

implementation of mathematical methods to correct both dependencies have been implemented within this system [13-15].

The main goal of this study is to clinically validate IViST as a new multi-sensor-based dosimetry platform for real-time plan monitoring in HDR brachytherapy.

2. IViST: the platform

IViST is a novel platform composed of 3 components: 1) a mPSD, 2) a light collection system, and 3) a Python-based graphical user interface (GUI) used for system management and signal processing. IViST can simultaneously measure dose, triangulate source position, and measure dwell time.

2.1 Dosimetry system

The optimized dosimetry system consisted of a 1.0 mm-diameter core mPSD (using BCF-60, BCF-12, and BCF-10 scintillators) coupled to 15 m-long fiber-optic cable EskaGH-4001 from Mitsubishi Rayon Co., Ltd. (Tokyo, Japan). The mPSD fiber was connected to a data acquisition system consisting of a compact assembly of photomultiplier tubes (PMTs) coupled with dichroic mirrors and filters to achieve a highly sensitive scintillation light collection [15]. The scintillation light generated inside the mPSD is collected in real-time and decoupled using multiple spectral bands [14, 15]. The PMTs readout is independent from the irradiation unit. Signal acquisition is made at a rate of 100 kHz using a data acquisition board (DAQ) type DAQ NI USB-6216M Series Multifunction from National Instruments (Austin, USA).

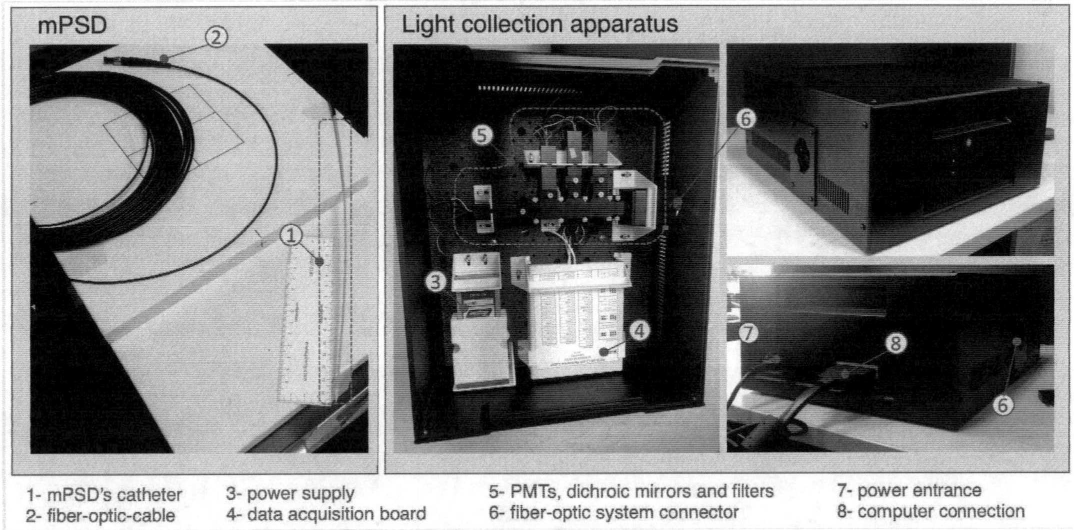

Figure 1. IViST's dosimetry system. A detailed list of sub-components can be found in Linares Rosales *et al* [16].

The design of the in vivo dosimetry system is shown in figure 1. The detector was made light-tight to avoid environ-mental light. The mPSD can be inserted into a 24-cm plastic catheter (Best Medical International, Springfield, VA, USA). In vivo measurements can be made using a supplemental catheter during HDR brachytherapy. The light collection apparatus was enclosed into a plastic custom-made box (43.2 x 38.1 x 18.42 cm^3) containing the light acquisition system as well as the DAQ. It is a passive light collection system that does not represent any electrical risk. The system's dosimetric performance have been evaluated in HDR brachytherapy, covering a range of 10 cm of source movement around the sensors [15,16]. Within a range of distances of 6 cm from the source, IViST can track the192Irsource with at least a 1 mm positional accuracy, and dose agreement with TG43-U1 [17] expected dose within

5 % [16].The detector further exhibited no angular dependence and the system high collection rate allows for dwell time assessment with a maximum average weighted deviation of 0.33±0.37s [16].

2.2 Graphical user interface for dosimetry system control

Figure 2 illustrates the graphical user interface (GUI) of the key software tools available to control the dosimetry system. The GUI was designed for simple operation by any users. PyQt5 was used to create the GUI. Figure 3 summarizes the main functionalities of the GUI and the relationship among them. The software's main windows give the user access to2 main modes: Offline and In vivo.

Within **Offline tools**:

- **Signal processing:** A measurement file is loaded in this module by the user. Then, the module allows for easy extraction and manipulation of the raw data previously collected. It can also be used to extract the statistics associated with each dwell position (e.g. measured dwell time, mean raw signal, mean dose per dwell).
- **System calibration:** The module receives a measurement data set, and based on TG43's expected dose, generates the calibration models for raw signal translation into a dose value. Among the main features of this module, it allows for dose prediction with the hyper-spectral approach proposed in [14] as well as a Random Forest algorithm and Neural Network
- **Dosimetry tools:** Performs the dosimetric analysis of the collected data. The user could, for instance, verify the agreement of the measured dose with that expected from TG43 and calculate the source position for the measured plan. The module also includes a catheter shift correction feature.

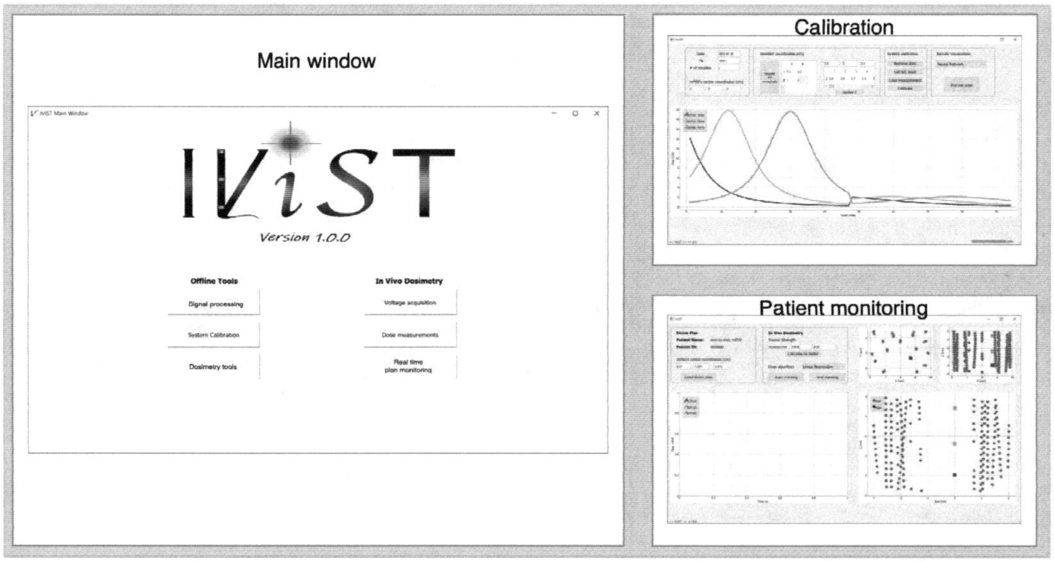

Figure 2. IViST's graphical user interface. Illustration of some of the available functionalities.

The In Vivo Dosimetry module contains the tools used for real-time data acquisition:

- **Voltage acquisition:** Collects the raw measurement data from each PMT in volts.
- **Dose measurement** is a similar to the offline module, excepting that a calibration data file has to be available in advance.

- **Real time plan monitoring:** Performs real-time tracking of a patient plan created in Oncentra Prostate and Oncentra Brachy (Nucletron, Elekta, Sweden). The module reads the plan data contained in the *.cha or DICOM RT files exported from the TPS, sets the treatment data as reference and monitors the patient plan in real-time. The system should be calibrated in advance. The GUI allows for real-time visualization of the measured dose at the mPSD location as well as the source location.

3. In vivo measurements for a prostate treatment: First Patient Measurements

In vivo dose monitoring was carried out for the first patient enrolled in an IRB approved protocol. Seventeen treatment catheters were positioned and reconstructed to deliver 15 Gy in a single fraction. An additional catheter devoted to measurements with the mPSD was introduced in the prostate volume. The mPSD's catheter was also reconstructed, but it was not used during the dose optimization process in the treatment planning system.

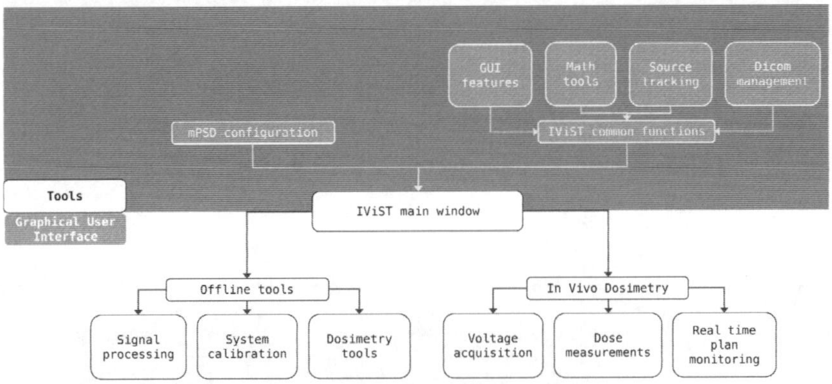

Figure 3. IViST's codes diagrams.

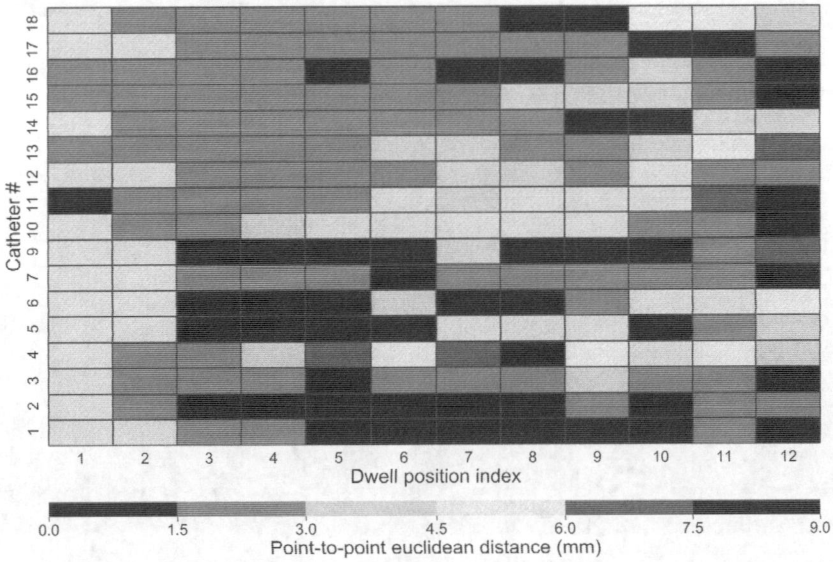

Figure 4. Source position triangulation results. Heat map of deviations between the planned position and the triangulated one in a 3D space. The cells in gray correspond to non-programmed dwell positions.

Figure 4 shows the heat map of deviations obtained be-tween the planned source dwell positions and the results from the triangulation with the mPSD. The x-axis corresponds to the dwell position index within each catheter, while the y-axis corresponds to the treatment catheter number. Catheter 8 is not plotted because it corresponds to the catheter dedicated to measurements with the mPSD. In total, 168 source dwell positions were tracked. As could be observed in figure 4, in most of the cases, the radial deviations remain below 3mm. However, there are some dwell positions with deviations reaching 9 mm. This effect could be explained by the small source dwell times, 0.1 s (from the TPS dose optimization process), programmed at these dwell-positions. Figure 5 presents the observed radial deviations as a function of the delivered dwell times for BCF10, BCF12, BCF60 sensors as well as for the mPSD's triangulation results. As could be observed in figure 5, for dwell times of 0.1 s, radial deviations in the order of 6 mm should be expected. IViST's data is collected at a rate of 100 kHz but averaged each 0.1 s. Thus a dwell time of 0.1 s makes it almost impossible to distinguish one single dwell position from the previous/subsequent one, leading to a failed source position triangulation for those locations. However, for any dwell-time of 0.5 s or more (which are the majority of the dwell times used in a prostate HDR treatment), the weighted average for source position triangulation is 2.17±1.15 mm.

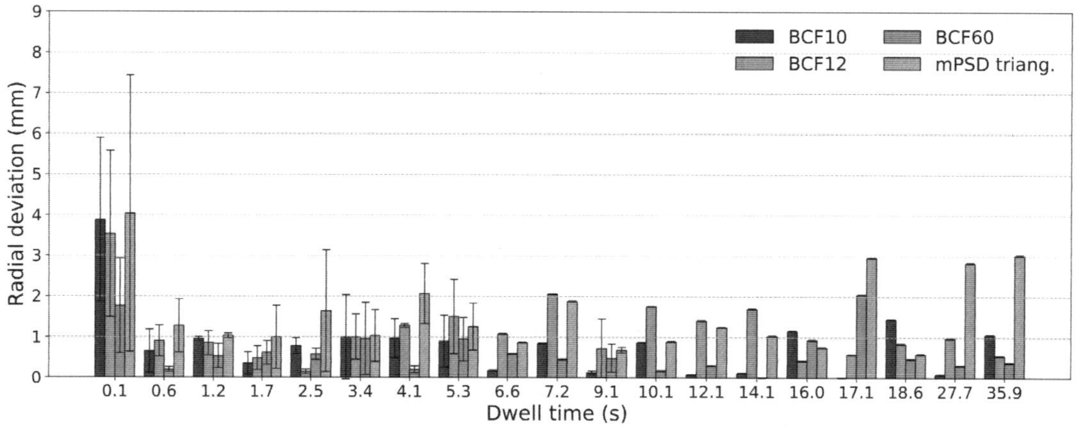

Figure 5. Scintillator deviation of the expected source-to-dosimeter radial deviations as a function of the source planned dwell time.

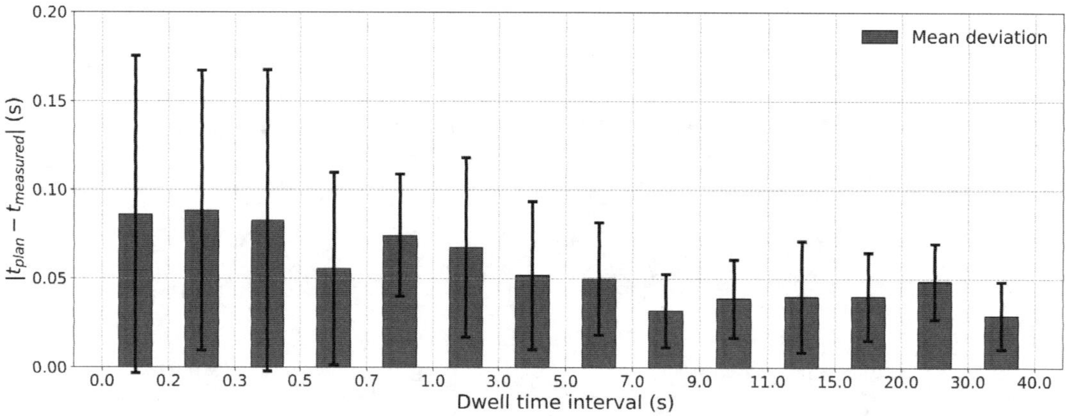

Figure 6. Deviation of mPSD measured dwell times from planned dwell times as function of planned dwell-time intervals.

Figure 6 shows the mean deviations between the measured dwell times for our mPSD system and the planned ones. The measured dwell times at all the distances to the source were agreement with the planned dwell time to better than 0.17 s. The overall average difference obtained was 0.06±0.04 s. Larger deviations were observed for dwell positions with dwell times inferior to 0.5 s, especially for those positions at long distances from the scintillators' effective volumes. Even though the longest mPSD-to-source distance measured for this clinical case was 55 mm, the high gradient field characteristic of the 192Ir source allowed to us obtain a sharp pulse of signal and, as a consequence, proper differentiation of the signal from one position to the subsequent one. A similar dwell time analysis was performed by Linares Rosales [16] et al. through in-water measurements during HDR brachytherapy. They reported a maximum deviation of 0.33±0.37 s for a range of mPSD-to-source distance up to 10 cm. However, for the same range of distance to the explored in this paper, the results are equivalents [16].

4. Conclusions

A novel, multi-sensor dosimetric system was optimized for HDR brachytherapy in vivo dosimetry and enabled measurements over a wide range of clinically relevant dose rate. The system presented allows real-time dose, source position and dwell-time measurements. It has numerous potential in vivo QA/QC applications beyond the currently available commercial dosimeters and build on its widely-known intrinsic energy independence and water equivalence properties. The first clinical in vivo case measured with iViST has been presented here and the clinical trial is on-going.

5. Acknowledgements

This work was supported by the National Sciences and Engineering Research Council of Canada (NSERC) via the NSERC-Elekta Industrial Research Chair grant # 484144-15and RGPIN-2019-05038. Haydee Maria Linares Rosales also acknowledges support from Fonds de Recherche du Quebec -Nature et Technologies (FRQ-NT).

6. References

[1] Fonseca G P, Podesta M, Bellezzo M *et al* 2017 *Phys. Med. Biol.* **62** 5440
[2] Cartwright L E, Suchowerska N, Yin Y *et al* 2010. *Med. Phys.* 37 2247-55
[3] Hardcastle N, Cutajar D L, Metcalfe P E *et al* 2010 *Phys. Med. Bi*ol. **55** 3859
[4] Kertzscher G, Andersen C E, Siebert F A *et al* 2011 *Radioth. Oncol.* **100** 456-62
[5] Seymour E L, Downes S J, Fogarty G *et al* 2011 *Med. Phys.* **38** 4785-94
[6] Kertzscher G, Andersen C E, Tanderup K 2014 *Med. Phys.* **41** 052102
[7] Johansen J G, Rylander S, Buus S *et al* 2018 *Brachytherapy* **17** 122-32
[8] Sethi R, Chun Kuo Y, Edraki B *et al* 2018 *Brachytherapy* **17** S114-5
[9] Beaulieu L, Beddar S 2016 *Phys. Med. Biol.* **61** R305
[10] Beddar S, Beaulieu L (Eds.) 2016 *Scintillation Dosimetry*. CRC Press. ISBN: 978-1-4822-0900-6
[11] Wootton L, Beddar S 2013 *Phys. Med. Biol.* **58** 2955
[12] Beddar S 2012 *Med. Phys.* **39** 6522-22
[13] Archambault L, Arsenault J, Gingras L *et al* 2005 *Med. Phys.* **32** 2271-8
[14] Archambault L, Therriault-Proulx F, Beddar S *et al* 2012 *Phys. Med. Biol.* **57** 7133
[15] Linares Rosales H M, Duguay-Drouin P, Archambault L *et al* 2019 *Med. Phys.* **46** 2412-21
[16] Linares Rosales H M, Archambault L, Beddar S *et al* 2020 *Med. Phys.* **47** 4477-90
[17] Rivard M J, Coursey B M, DeWerd *et al* 2004 *Med. Phys.* **31** 633-74

Further investigation of lung tumour peripheral doses using normoxic polymer gel dosimetry techniques

A Venning[1], M Mundayadan Chandroth[1], C Morgan[2] and M Roberts[2]
[1]Mid-North Coast Cancer Institute, Port Macquarie Base Hospital, Port Macquarie, NSW, Australia
[2]Mid-North Coast Diagnostic Imaging, Highfields Circuit, Port Macquarie, NSW, Australia

Email: anthony.venning@health.nsw.gov.au

Abstract. This work builds upon previous investigations of lung tumour peripheral doses for 6 MV, 6 MV FFF, and 10 MV FFF conformal arc therapy beams calculated in the Monaco® TPS and delivered using an Elekta® Agility™ linac. An improved patient lung phantom is developed with measurements using the normoxic PAG gel dosimeter and compared against dose planes from the TPS. The gel dosimeter measurements indicate that the TPS is overestimating the secondary build-up in the lung tumour peripheral region. It has been determined that in lung tumours, 6 MV FFF is the optimal beam energy for peripheral dose coverage and that there is a dosimetric compromise using 10 MV FFF.

1. Introduction

In recent years, radiation therapy has made significant technological advances with the introduction of treatment techniques like volumetric modulated arc therapy (VMAT), stereotactic radiotherapy (SRT), and flattening filter free beams (FFF). These treatment techniques, while enhancing patient outcomes and care [1-3], come with significant consequences if misadministration occurs. Accurately determining delivered dose within tight margins to an inhomogeneous treatment volume, as in lung tumours, using complex delivery techniques, is extremely difficult and demanding. Therefore, a high level of confidence needs to be placed in the ability of the treatment planning system (TPS) to correctly calculate both dosimetrically and geometrically within the patient volume. For high dose fractionation regimes, beam geometry is of high importance, and as such, the delivery time is an important factor for a non-static treatment site, like the lung. With longer delivery times, geometric errors increase due to patient motion. At our treatment centre, radiation oncologists prefer to use 10 MV FFF due to its higher dose rate compared to 6 MV and 6 MV FFF, to minimise treatment time and motion during delivery. Clinical physicists are concerned though, that the higher energy extends the secondary build-up region in lung tumours, reducing the dose to the tumour periphery [4]. This is an important factor to consider, as an under dose rapidly leads to a loss of tumour control. This phenomenon was previously investigated [5] for the photon beams available in our clinic, and this current work utilises the unique, tissue equivalent, high-resolution properties [6,7] of gel

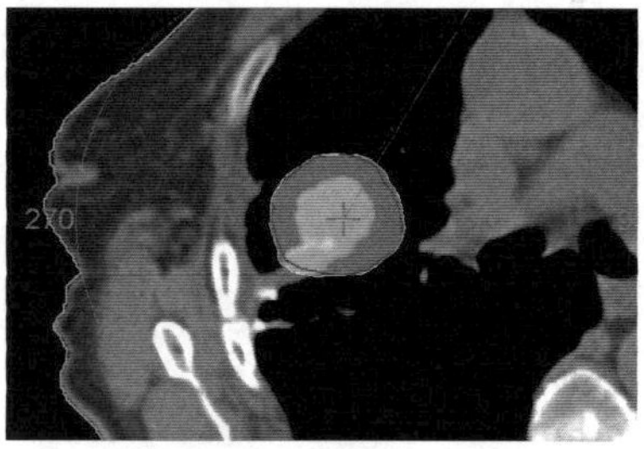

Figure 1. PTV in lung.

dosimeters to build upon previous findings using an improved phantom design.

2. Methods and Materials

2.1. Lung-tumour phantom

The phantom was created from foam and cut to the size of a typical human thorax. A hole of 47 mm dia. was drilled out of the foam centre to hold the dosimeter. The foam was wrapped in 1 cm thick Bolus (NL-Tec Pty Ltd) to simulate the chest wall. The foam was soaked in water with the correct retention determined to be equivalent to that of the electron density of lung. Plastic (PET) vials of 47 mm dia. x 150 mm hgt. filled with 5% gel and 95% water were inserted into the foam as shown in figure 2.

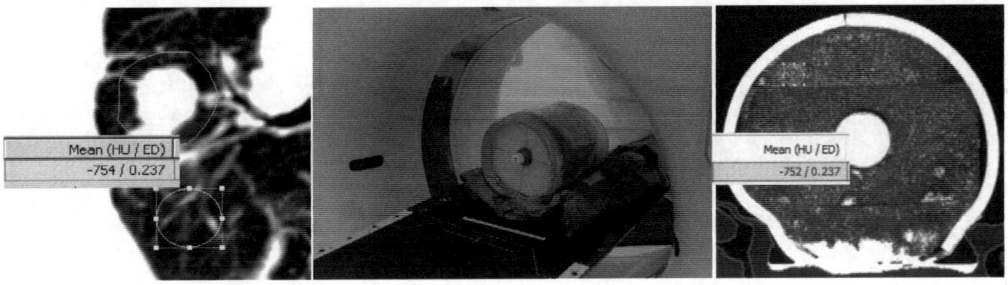

Figure 2. CT of lung-tumour phantom showing the equivalent lung electron density.

2.2. Gel dosimeter manufacture

Manufacturing of the normoxic PAG dosimeter is well described elsewhere [8-16]. The composition used was 5% gel, 89% de-ionised water, 3% acrylamide, 3% N, N′-Methylene-bis-acrylamide, 2 mM hydroquinone and 5 mM tetrakis phosphonium chloride. Seven normoxic PAG gel filled vials were used in this study, 3 for calibration, 3 for lung tumour irradiation and 1 for PDD simulation.

2.3 Treatment planning

The lung phantom with gel insert was CT scanned and the data set sent to the TPS. Conformal arc beams of 360° were created using 6 MV, 6 MV FFF and 10 MV FFF beam energies, all with a field size of 4.8 cm^2, 450 MUs and calculated at 1%/1 mm resolution. The transverse central axis dose plane was exported for comparison with the gel dosimeter measurement.

2.4 Gel irradiation

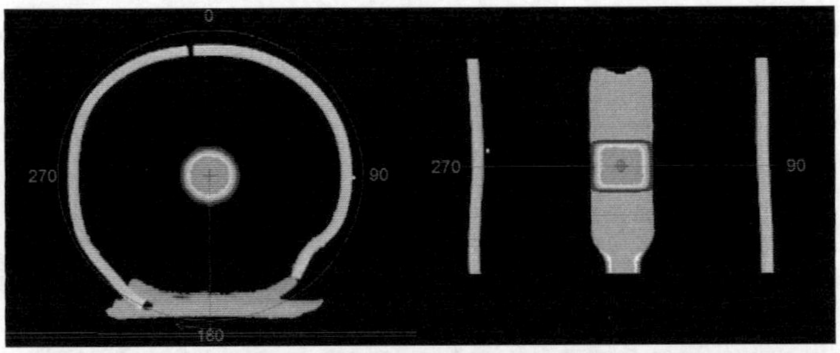

Figure 3. Conformal arc treatment plan in Monaco.

The lung phantom was placed on the treatment couch, a cone beam CT was performed and then position matched to the reference data using HexaPOD™. Gel filled vials were carefully inserted and irradiated with the conformal arc plans. The calibration curve was generated using the PDD method in water. One vial was irradiated side-on at the surface with 6 MV to assess the dosimeter in the build-up region compared to the TPS and beam reference data.

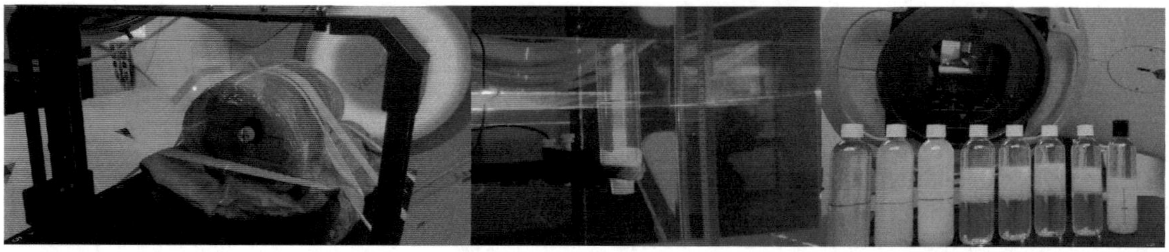

Figure 4. Gel dosimeter irradiation.

2.5 MRI readout

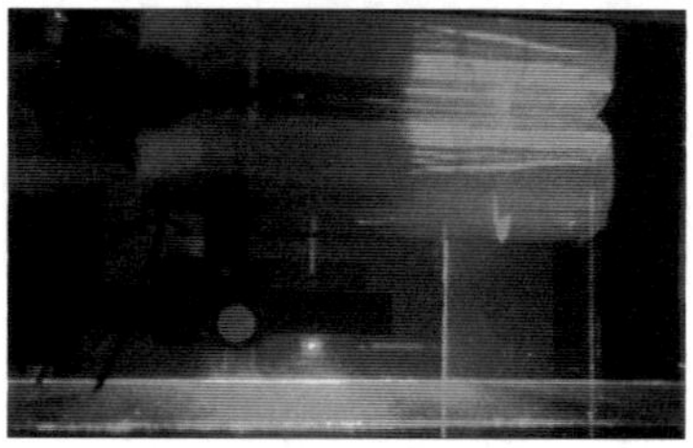

Figure 5. PDD irradiation.

Five days after irradiation, the gels were imaged using a clinical 3T MRI scanner in a head coil to enhance the signal-to-noise ratio. To generate T_2 maps, a multi-echo sequence with imaging parameters; echoes = 16, TE = 40 ms, FOV = 128 mm^2, matrix size 128 mm^2 and slice thickness of 5 mm was used. The vials were imaged on the central transverse plane for the open field 360° arc deliveries and at various known dose deposition positions along the calibration vials.

2.6 Beam profile analysis
The MRI images were processed and compared against the TPS using MATLAB™ [17]. T_2 maps were converted to dose maps; however, the profiles were normalised to match the central region of the TPS profiles so an effective assessment of the peripheral build-up region could be made.

3. Results and Discussion

3.1. R_2-dose response curve

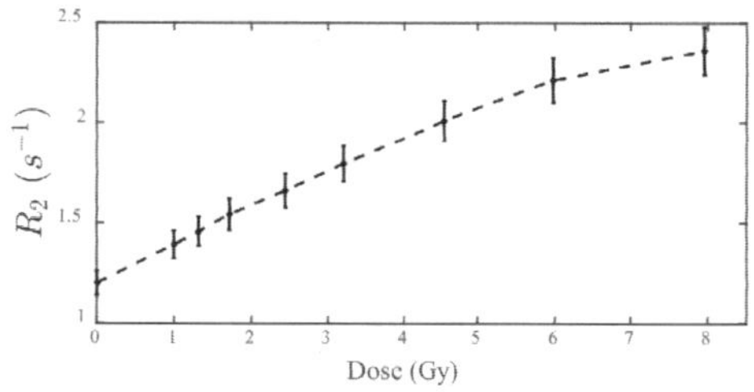

Figure 6

The R_2-dose response curve in figure 6 has an R_2-dose sensitivity of 0.19 s^{-1}Gy^{-1} in the linear region.

3.2. Gel measurements versus TPS
Figure 7 shows the dose distribution for the three different beam energies and it is shown that for equivalent MUs there is higher dose deposition for 6 MV FFF. Figure 8 shows profile comparisons of the dose maps, indicating the TPS is overestimating the dose in the tumour peripheral region, which agrees with previous findings [5]. This is concerning considering the requirement of the tight dose margins when using high dose fractionation regimes.

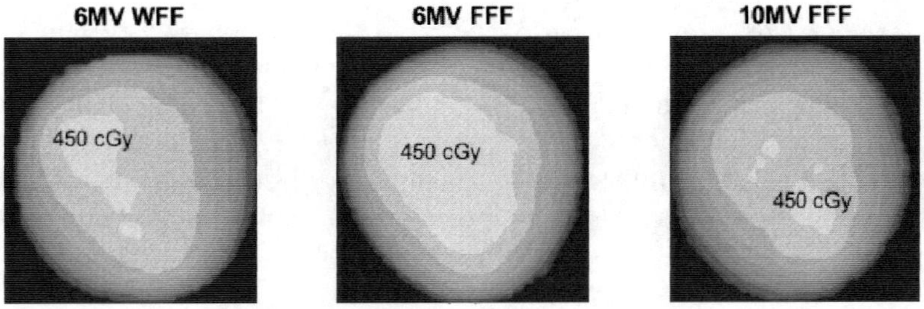

Figure 7. Gel dosimeter dose maps.

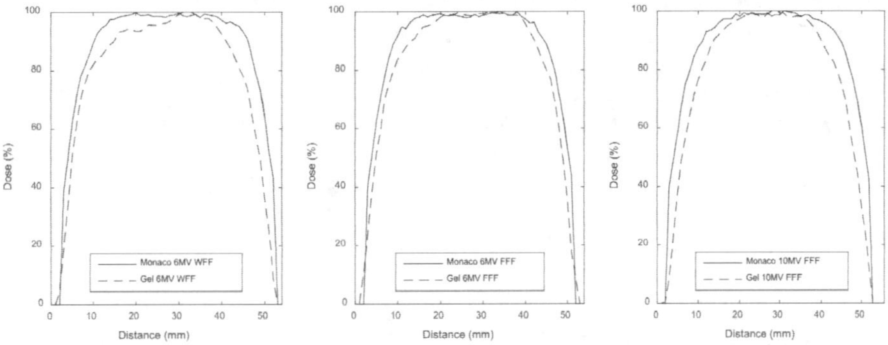

Figure 8. Monaco versus gel dosimeter profiles.

To verify the gel dosimeters correctly measure the build-up region, the gel PDD was plotted against the reference beam data and also compared against the TPS as shown in the left plot in figure 9. These results show an underestimation between gel dosimeter and ion chamber (5% at the surface) and Monaco (10% at the surface), which contrasts with the lung tumour profiles in figure 8 above.

3.3. Energy evaluation in lung tumours

A comparison of the dose profiles is shown in figure 9 for each energy used in the gel dosimeters. The 6 MV FFF beam shows the higher absorbed dose while 10 MV FFF shows a lower absorbed dose for equivalent MUs. 6 MV FFF also shows a higher dose deposition at the tumour periphery. While 6 MV also shows a higher peripheral dose compared to 10 MV FFF, the authors suggest that 6 MV FFF is the optimal energy to use for lung tumours, providing greater coverage and peripheral dose. If using 10 MV FFF in lung, a higher total dose needs to be delivered to expand the Rx iso-dose line, potentially compromising organs at risk and therefore, careful consideration should be given to its use.

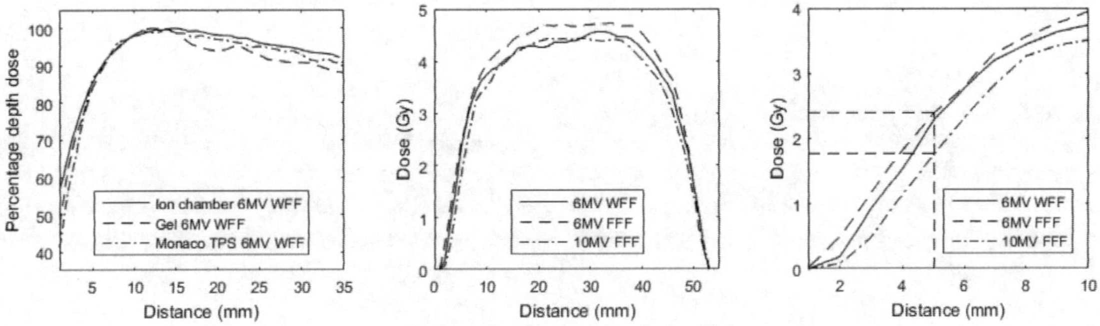

Figure 9. Overlaid gel profiles.

Journal of Physics: Conference Series **2167** (2022) 012025 doi:10.1088/1742-6596/2167/1/012025

4. Conclusion

This study builds upon work previously performed by the lead authors by developing a lung phantom that improves patient inhomogeneity lung simulation. Further to the previous study, the build-up region of the gel dosimeter has been examined for a 6 MV beam and found to closely represent ion chamber measurements. It was also found that the Monaco TPS is over estimating the peripheral lung tumour dose when compared with gel dosimeters. The recommended beam energy in lung tumours is 6 MV FFF followed by 6 MV. Using 10 MV FFF in lung tumours requires careful consideration, as there is found to be a dosimetric compromise compared with 6 MV and 6 MV FFF.

5. References

[1] Prendergast B *et al* 2012 *J. Appl. Clin. Med. Phys.* **14** 64-71
[2] Xing L *et al 2005 Med. Dosim.* **31** 91-112
[3] Folkert M and Timmerman R 2017 *Adv. Drug Del. Rev.* **109** 3-14
[4] Wang L *et al* 2002 *J. Appl. Clin. Med. Phys.* **3** 51-9
[5] Venning et al 2019 *J. Phys.: Conf. Ser* **1305** 012003
[6] Baldock C *et al* 2010 *Phys. Med. Biol.* **55** R1-R63
[7] Venning A J *et al* 2005 *Med. Phys.* **32** 1047-53
[8] Venning A J *et al* 2004 *J. Phys.: Conf. Ser.* **3** 155-8
[9] Venning A J *et al* 2005 *Phys. Med. Biol.* **50** 3875-88
[10] Hill B *et al* 2002 *Phys. Med. Biol.* **47** 4233
[11] De Deene Y *et al* 2002 *Phys. Med. Biol.* **47** 2459-70
[12] De Deene Y *et al* 2002 *Phys. Med. Biol.* **47** 3441-63
[13] De Deene Y 2004 *J. Phys.: Conf. Ser.* **3** 34-57
[14] Baldock C *et al* 1998 *Phys. Med. Biol.* **43** 695-702
[15] De Deene Y *et al* 2006 *Phys. Med. Biol.* **51** 653-73
[16] De Deene Y and Baldock C 2002 *Phys. Med. Biol.* **47** 3117-41
[17] Murry P *et al* 2000 *Austral. Phys. Eng. Sci. Med.* **23** 44-51

RSC: Optically stimulated emission of LiF:Mg,Cu,P - towards 3D optically stimulated luminescence dosimetry

Jacob S Nyemann[1], Mads L Jensen[1], Camilla L Nielsen[1], Karl-Emil T Bondgård[2], Rosana M Turtos[1], Brian Julsgaard[1,2], Jørgen B B Petersen[4], Ludvig P Muren[3,4] and Peter Balling[1,2]

[1]Department of Physics and Astronomy, Aarhus University, Denmark
[2]Interdisciplinary Nanoscience Center (iNANO), Aarhus University, Denmark
[3]Danish Centre for Particle Therapy, Aarhus University Hospital, Denmark
[4]Department of Medical Physics, Aarhus University and Aarhus University Hospital, Denmark

E-mail: jacob1soegaard@phys.au.dk

Abstract. With the introduction of highly conformal treatment modalities, dose verification in 3D is becoming more important than ever for patient-specific quality assurance of radiotherapy. Reusability of 3D dosimeters may be the path to cope with the cost-benefit issues caused by batch-to-batch fluctuations and intense calibration protocols in existing 3D systems. We present the idea of an envisioned (optically stimulated luminescence) OSL-based 3D readout system, which exploits the inherently reusable dosimetry properties of OSL. We provide the emission spectra of the OSL active material LiF:Mg,Cu,P (MCP) for three stimulation wavelengths (460 nm, 532 nm, and 664 nm), and summarize recently published optical characterization results to highlight the requirements of a readout system for an MCP-based dosimeter.

1. Introduction

Research in 3D dosimetry has been of great interest during the past decades [1-4]. The well-established need for 3D dosimetry becomes continuously more relevant as highly conformal radiotherapy modalities are utilized worldwide [2-7]. At the time of writing, multidimensional arrays of ion chambers and diodes are clinically used as semi-/pseudo-3D dosimeters, while true 3D dosimeters are mainly used at research institutes [3]. Thus, a widespread introduction of clinically accepted (true) 3D-dosimetry systems still lies ahead. Key elements delaying the implementation of true 3D dosimetry in quality-assurance routines are the high-cost, batch-to-batch fluctuation, calibration, and lack-of-reusability properties shared by most 3D-dosimetry systems. Our group has previously presented the idea of an (optically stimulated luminescence) OSL-based 3D dosimetry system to overcome the reusability issue [8, 9]. With a reusable dosimeter, batch-to-batch fluctuations and tedious calibration protocols become much less important.

The envisioned OSL-based 3D dosimetry read-out system is outlined in figure 1, where a laser source provides the stimulation light from the left. The laser beam is first collimated, before it is transformed into a sheet of laser light by the combination of a Powell lens and a Fresnel lens. The plane of laser light is used to stimulate the composite dosimeter material sheet by sheet, providing the

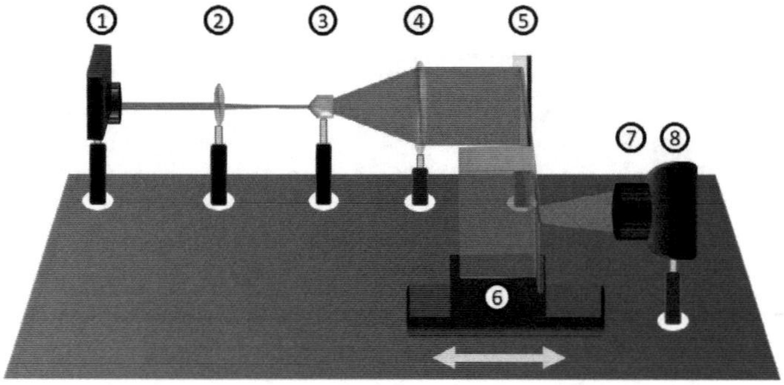

Figure 1. Schematic of a possible future 3D readout system. The setup contains: 1. blue laser, 2. collimating lens system, 3. Powell lens, 4. Fresnel lens, 5. mirror, 6. translation stage, 7. imaging optics and spectral filters, 8. CCD camera. The blue line/plane is the stimulating laser light, while the purple light (between 6 and 7) indicates the OSL emission captured by the camera.

third dimension by moving the dosimeter with a translation stage. The composite dosimeter material, placed on top of the translation stage, should consist of small particles of OSL-active material embedded inside a transparent silicone matrix as presented in [8]. The stimulation of OSL-active particles will yield a dose-dependent OSL emission. The use of imaging optics and spectral filters in front of a camera will allow the acquisition of a series of dose-dependent 2D images of the stimulated sheets of the 3D dosimeter.

A potential OSL material for the 3D dosimetry system presented in figure 1 is the near-tissue-equivalent LiF:Mg,Cu,P (MCP) with $Z_{\mathrm{eff}} = 8.2$. MCP is best known for its TL properties but has also been investigated for its OSL properties in recent years [10-13].

Here, we present the emission spectra of MCP for three different stimulation sources. We also summarize dose-response and emission-intensity results presented in [13]. Emission spectra allow the identification of necessary equipment for a readout system, e.g. spectral filters. Dose response and emission-intensity information provides an estimate of the measurable dose-region, where the upper limit is characterized by a small deviation from linearity. Such deviation can be found by a saturation curve fit of the type $S = S_0\left(1 - e^{-D/D_s}\right)$ [14-16], where D_s indicates a dose with ~ 37 % lower dose response than for a linear behaviour.

2. Materials and methods

MCP emission was measured under truncated-Gaussian-continuous-wave stimulation by 460 nm (blue), 532 nm (green), and 664 nm (red) lasers. Wavelength-dispersed emission was monitored by a CCD sensor (PIXIS 100BR) attached to the back of a spectrometer (Princeton Instruments Acton SP2358). Several wavelength regions were measured using a combination of spectral filters. All samples were irradiated by 20(1) Gy water-equivalent dose with a Cs-137 gamma source. Measurements were acquired 10 to 40 minutes after radiation.

3. Results and discussion

The emission of MCP shown in figure 2 shows the spectra under blue (460 nm), green (532 nm), and red (664 nm) stimulation respectively. The decaying OSL signal is seen between 340 nm and 400 nm for the blue stimulation, between ~ 400 nm and 470 nm for the green stimulation, while no decaying OSL emission is observed for red stimulation, as previously demonstrated [13]. It is observed that all three emission spectra show a steeply increasing anti-Stokes shifted signal from the pump laser with its wavelength located in the filtered region. All stimulation wavelengths show an intense emission in the region for wavelengths higher than the stimulation wavelength.

As noted by Nyemann et al., the lack of OSL emission under red stimulation indicates a room-temperature activation energy of the OSL active dosimetric state above 1.85 eV. Similarly, it must be lower than 2.33 eV due to the observation of OSL for the green stimulation. The steep anti-Stokes shifted emission observed for all three stimulation wavelengths can be interpreted as a so-called Raman continuum [13, 17, 18]. The spectra presented here further support this interpretation, as all

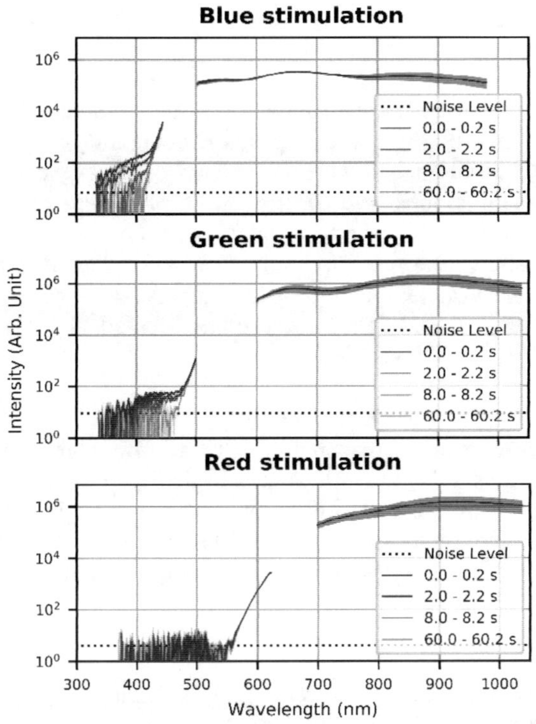

Figure 2. Emission from MCP stimulated by a blue, green, and red laser respectively. Each spectrum is shown at four distinct timespans and is given as the mean value of three measurements with different samples. The emission has been calibrated against filter and spectrometer response. The standard deviation of the three measurements is marked by the shaded area associated with each time span. The statistical fluctuations of the background (measured in the short-wavelength region) is termed "Noise level" and has been indicated by the black dotted line. Short-wavelength data in time span "0.0–0.2 s" and "60.0–60.2 s" are reproduced from [13].

three stimulation wavelengths produce somewhat similar characteristic emission (note that a direct comparison should be made on an energy scale). The long-wavelength regions (longer than the stimulation wavelength) are very similar for all stimulations with a clear indication of an emission band near 900 nm. The intensity of the long-wavelength emission emphasizes the need for efficient filtering of all other wavelengths than the OSL emission window, for any of the selected stimulation sources. It is also necessary to filter out the stimulation wavelength and the anti-Stokes shifted Raman continuum. For OSL readout with a blue stimulation source this means that spectral filters should be included at position 6 (see figure 1) to only allow transmission of emission below ~ 400 nm. Similarly, the lack of OSL under red stimulation puts a clear upper limit on the wavelength of the stimulating laser (position 1) for the 3D readout system.

The dose response for the OSL emission of MCP under blue stimulation (340-400 nm) is shown in figure 3, with the OSL decay for the 10 Gy dose in the inset. This dose response has a sublinear behaviour where a saturation fit yields a saturation dose of 75(12) Gy, which at 10 Gy corresponds to a 6 % deviation from linearity. This suggests that doses below 10 Gy for MCP can easily be monitored when using appropriate calibration procedures [13].

The dose response at 10 Gy was by Nyemann et al. used to estimate the isotropic photon flux emitted by a cubic millimeter of MCP per Gy to be $3.3 \cdot 10^8$ photons / (Gy·mm^3) in the full solid angle of 4π. This emission was used to estimate a minimum dose of approximately 0.1 Gy that can be measured with a S/N ratio of 50 for MCP particles embedded in a 3D-dosimeter at a 1 % concentration [13]. Assuming negligible contributions from systematic effects, the accuracy will be dominated by read-out precision, yielding an accuracy of a few percent. This shows that MCP has the potential to constitute the OSL-active material in a future 3D-dosimetry system with a measurable dose range of 0.1-10 Gy.

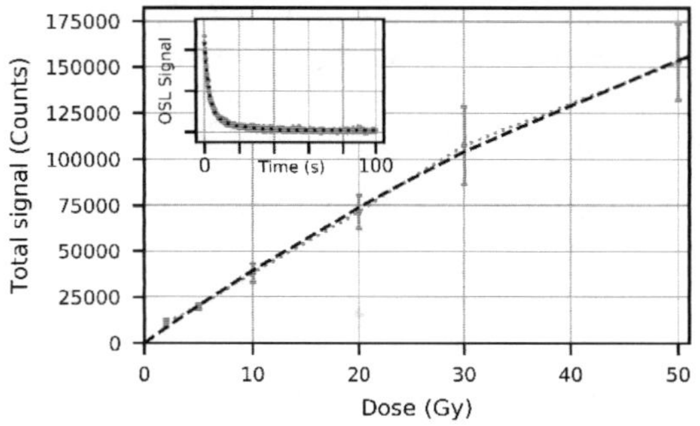

Figure 3. Dose–signal relation for the integrated OSL-emission (340–400 nm) over the first 20 seconds is shown as a mean value of 9 measurements (3 samples repeated 3 times). The error bars show the standard deviation on the 9 measurements. The black dashed line presents a saturation curve fit of the type given in the introduction, with $D_s = 75(12)$ Gy. The inset shows the time domain of the OSL signal at 10 Gy, where the dotted line represents a double-exponential fit. The figure has been adapted from [13].

4. Conclusion

Optically stimulated luminescence processes are shown to be a reasonable pathway toward reusable 3D dosimeters. The emission spectra of MCP are presented for three different stimulation sources. It is shown that a stimulating laser for an MCP-based system must have a wavelength below 664 nm, and that a system based on a 460 nm laser must use filters that block emission > 400 nm efficiently. The OSL active material MCP is estimated to provide a measurable dose range of 0.1–10 Gy for an MCP-silicone 3D dosimeter.

5. Acknowledgements

This research was supported by the Independent Research Fund Denmark | Technology and Production Sciences and the Novo Nordisk Foundation.

6. References

[1] Gore J C *et al* 1996 *Phys. Med. Biol.* **41** 2695-704
[2] Doran S J 2009 *Appl. Radiat. Isotopes* **67** 393-398
[3] Mijnheer B (ed) 2018 *Clinical 3D dosimetry in modern radiation therapy* (Boca Raton, Florida: CRC Press)
[4] Taño J *et al* 2019 *J. Phys. Conf. Ser.* **1305** 012034
[5] Høye E M *et al* 2015 *J. Phys. Conf. Ser.* **573** 012067
[6] Høye E M *et al* 2017 *J. Phys. Conf. Ser.* **847** 012021
[7] d'Errico F *et al* 2017 *Radiat. Meas.* **106** 612-617
[8] Sądel M *et al* 2017 *J. Phys. Conf. Ser.* **847** 012044
[9] Balling P *et al* 2018 *European Patent Application* EP3264137A1
[10] Bilski P *et al* 2014 *Radiat. Meas.* **71** 61-64
[11] Sądel M *et al* 2019 *Radiat. Meas.* **123** 58-62
[12] Sądel M *et al* 2019 *Radiat. Prot. Dosim.* Advance online publication 1–8
[13] Nyemann J S *et al* 2020 *Radiat. Meas.* **138** 106390
[14] Kristianpoller N *et al* 1974 *J. Phys. D Appl. Phys.* **7** 1063-1072
[15] Sunta C M *et al* 1994 *J. Phys. D Appl. Phys.* **27** 852-860
[16] Yukihara E G and McKeever S W S 2011 *Optically Stimulated Luminescence: Fundamentals and Applications* (Chichester, West Sussex: John Wiley & Sons)
[17] Godfrey-Smith D I *et al* 1989 *J. Lumin.* **44** 47-57
[18] Huntley D J *et al* 1989 *J. Lumin.* **44** 41-46

3D dose delivery QA using couch and gantry mounted cameras

M. Ramish Ashraf [1], Cedar Farwell [2], Daniel A Alexander[1], Rongxiao Zhang[1,3,4], David J Gladstone[1,3,4], Brian W Pogue[1,2,4] and Petr Bruza[1,2]
[1]Thayer School of Engineering, Dartmouth College, Hanover NH 03755, US
[2]DoseOptics, Lebanon, NH 03756, USA
[3]Department of Medicine, Geisel School of Medicine, Dartmouth College Hanover NH 03755 USA
[4]Norris Cotton Cancer Center, Dartmouth-Hitchcock Medical Center, Lebanon, NH 03756 USA

Email: petr.bruza@dartmouth.edu

Abstract. While 2D planar radiation therapy quality assurance techniques suffice for linac verification, a full 3D dosimetry measurement is ideal for robust treatment plan verification. In this study, an optical 3D dose reconstruction technique based on radioluminescence imaging is presented, using two orthogonal views provided by gantry and couch mounted cameras. Both small field and multi-target beam apertures were tested, mimicking typical 3D radiosurgery plans. Pass rates >94% for 3D gamma with 2%/2mm were achieved in all cases. Slice images were used for gamma analysis in each of transverse, sagittal and coronal planes, through the point of maximum dose. This proposed system can enable real-time imaging of beam delivery dynamics in a cylindrical water-equivalent phantom with high spatial resolution 3D reconstruction of the dose distribution, and verification of each beamlet used.

1. Introduction

External beam radiation therapy relies upon intensity and spatial modulation of the beam to deliver dose distributions, which are highly conformal and exhibit steep gradients. It is imperative that prior to complex patient treatments, the plans be tested in a phantom with the dose distribution measured and compared to the treatment planning system (TPS) for quality assurance (QA). Any dose delivery abnormalities can then be caught prior to patient irradiation. Currently, patient specific QA is commonly performed with 2D planar measurements acquired via electronic portal imaging device (EPID), diode arrays, and ion chamber arrays. However, most charge based dosimeters such as diodes and ionization chambers suffer from coarse spatial resolution and dose-rate dependence. Additionally, with the increased usage of 3D imaging modalities, tumor volume and organs at risk (OAR) delineation is now routinely performed on multiple cross sectional slices. Therefore, a dosimeter which can provide 3D dose information is an attractive QA tool because it can potentially be used to assess where exactly the treatment plan fails and if the resulting failure would translate into clinically meaningful concerns. In this study, a high resolution dosimetry imaging solution is tested in a water tank. Organic liquid scintillator solutions, due to their ideal dosimetric properties lend themselves as an attractive tool for 3D dosimetry. Although traditionally, scintillation and Cherenkov based detectors are used either as point detectors or as 2D planar QA devices when used in conjunction with an optical camera or optical fibers [1, 2], there have been advances in recent times in terms of 3D dose characterization [3, 4]. In particular, Bruza *et al* [5] recently showed that combining EPID with optical imaging of radioluminescence can be used to reconstruct dose in 3D in real-time. In this study, we extend the reconstruction technique first shown by *Bruza et al* by incorporating an additional gantry mounted camera which yields a measurement that is completely optical in nature and is devoid of scattering and non-water equivalence issues typically encountered in EPID. The technique is presented as a proof of concept in this study.

2. Methods and Materials

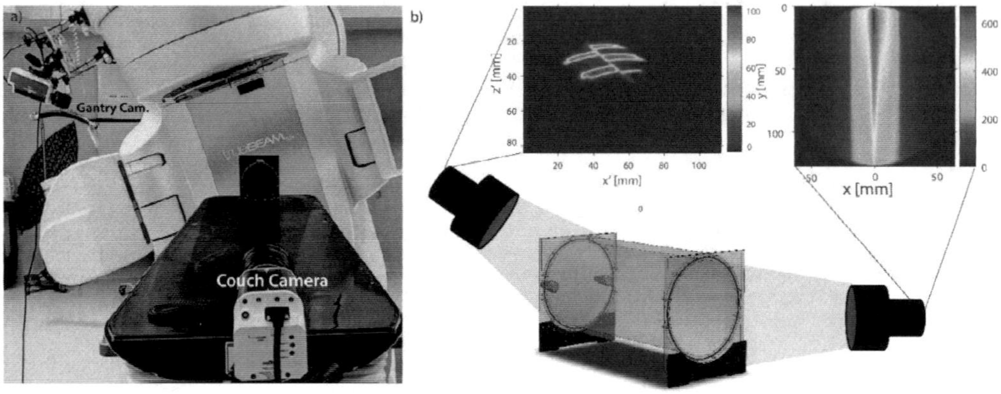

Figure 1. Experimental setup showing the two camera setup. The gantry camera images a scintillating sheet wrapped around the cylindrical phantom, while the couch camera images the projected depth dose information from the front imaging face.

Two blue sensitive intensified cameras (C-Dose, DoseOptics LLC., Lebanon, NH, USA), were gated to X-ray pulses produced by a TrueBeam (Varian Medical Systems, Palo Alto, CA) linear accelerator. The gantry mounted camera was equipped with a 50 mm/F/2 lens, while the couch mounted camera was equipped with a 135 mm/F/4 lens (Canon, Ota City, Tokyo, Japan). A 6 MV static 3 x 3 cm^2 beam and a static MLC defined apertures simulating a beam aperture for a typical control point in a multi-target SRS plan were delivered to the phantom with a total beam delivery of 100 Monitor Units (MU), where 1 MU is defined as 1 cGy at a depth of maximum dose for a reference 10 x 10 cm^2 field. The cameras acquired at frame rate of 20 fps, corresponding to 0.5 (MU) per frame for maximum pulse rate of 6 MV X-ray beam. A cylindrical acrylic phantom containing a radioluminescent solution with diameter 13 cm and length 30 cm was used as the dose deposition medium. The phantom was filled with distilled water and doped with the fluorophore/scintillator Quinine Sulfate (0.5g/L) (99% Quinine Hemisulfate monohydrate, Sigma Aldrich, St Louis, MO). The setup is shown in Figure 1 which depicts the two cameras imaging the radioluminescent phantom. The phantom was blacked out from the inside except the front imaging face. Therefore, to facilitate imaging of the surface of the cylinder, a blue emitting scintillating sheet (Gd_2O_2S: Tb, PJ X-ray, New Jersey, USA) was wrapped around it. The sheet was 0.2 mm thick, therefore no appreciable beam perturbance is expected for high energy x-rays. To transform the gantry camera image from a curved surface to a planar one, a checkerboard wrapped around the curved surface of the cylinder was imaged. A polynomial transformation of order 4 was then applied to 'unwrap' the raw images to a planar surface, yielding a beam's eye view of the beam aperture. Both set of images (couch and gantry cameras) were background subtracted and flatfield corrected to account for optical vignetting. Additionally, due to the curved nature of the surface, the half of the cylinder which is distal from the gantry camera would result in lower intensity being captured at the sensor plane compared to half that is proximal to the camera. To account for this, a 7 x 7 cm^2 square field was imaged in the same geometry and used as a secondary flatfield correction for the gantry camera images. A schematic of the experimental setup and the realization is shown in Figure 1. Figure 1 b) shows the images acquired with the two cameras for one of the complex apertures used in this study. A 3D volume is reconstructed using information provided by the two orthogonal views. The reconstruction method has been previously used to reconstruct 3D dose volumes using a single optical camera and additional information provided via delivery log files or EPID, both of which provide orthogonal view to the couch mounted camera.

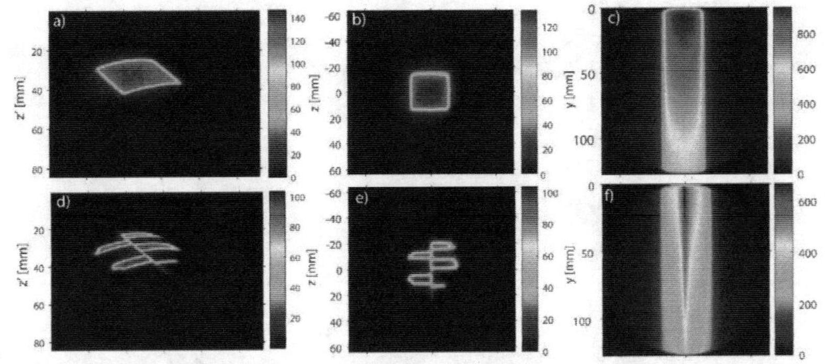

Figure 2. a) Image of the gantry camera view for the 3 cm^2 beam. b) Beam's eye view obtained from the image in a) after applying a polynomial (order 4) transformation. c) couch camera view of the 3 cm^2 square beam. d), e), f) show the same image sequence as described above but for the static multitarget control point.

The main reconstruction steps are briefly explained again below:

1) The transformed gantry image is back projected through the volume of the phantom.
2) The back projected image is then summed along the optical axis, which yields an image similar to the one obtained by the couch mounted camera.
3) A convolution kernel relating the two projected images is then found and assuming a radially symmetric kernel because of the homogenous composition of the phantom, the kernel was then extended to the 3rd dimension using the inverse Abel transform.
4) The kernel is then convolved with the back projected volume which yields the reconstructed 3D dose volume inside the phantom.

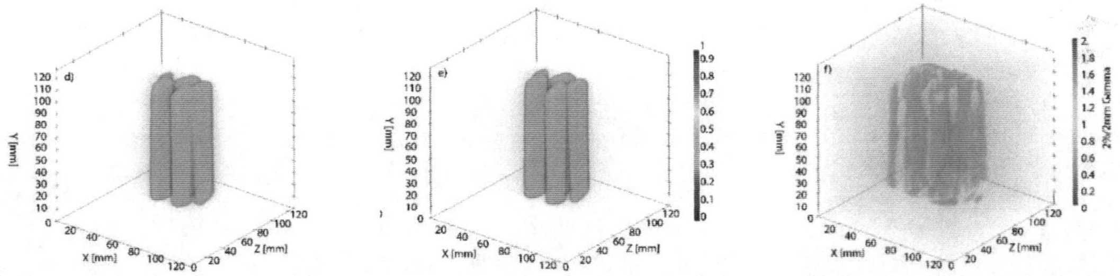

Figure 3. The optical 3D dose reconstruction is shown for one of the multitarget fields in a) and Simulated 3D dose distributions obtained from the treatment planning system for the same fields are shown in b). The 2%/2mm gamma index with global normalization is shown in c). The 3D passing rates for the three beams was 97%. Global dose normalization to the point of maximum dose was used along with a 5% dose threshold.

The reconstructed optical data was then compared to the simulated dose distribution obtained through the treatment planning system (TPS) via 3D and 2D slice gamma index analysis.

3. Results

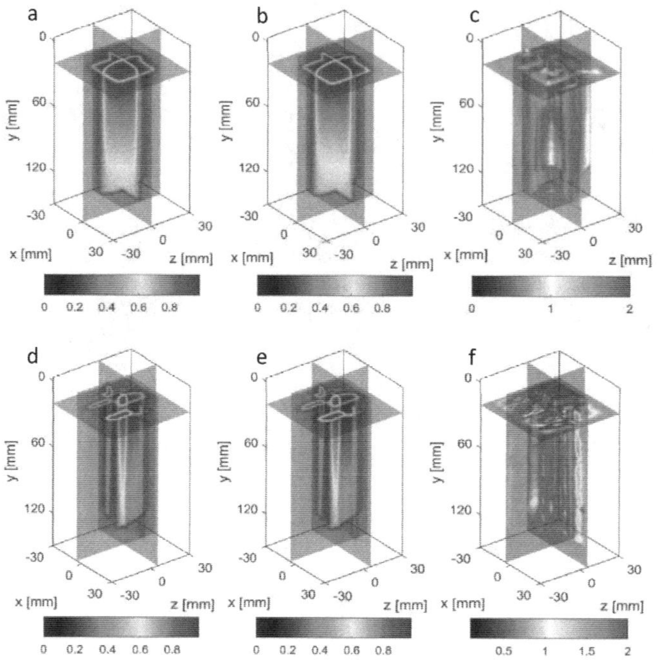

Figure 4. a) and d) represent the transverse, sagittal and coronal planes are shown at the point of maximum dose for the optical data for the square beam and the multitarget aperture, respectively. In b), e) the same planes are shown, but for the TPS distribution. In c) and f) the 2%/2mm gamma index map is shown for each slice. The pass rates for these criteria were 99% and 96%, respectively.

Figure 2 depicts the images acquired for the square beam and one of the multitarget beam apertures. The second column shows the gantry camera view after the applying the polynomial transformation and correcting for the curvature of the phantom. The third column shows the image acquired with the couch camera. The result for the 3D dose reconstruction of the first multitarget beam is shown in Figure 3. A passing rate of 97 % was observed for the static multitarget beam. Global normalization to the point of maximum dose and a dose threshold of 5% was used in all cases. When the gamma criterion was increased to 3%/3mm, all beams exhibited a >99% passing rate. Additionally, 2D slice gamma was performed for three orthogonal planes at the point of maximum dose for the multitarget beam. The results are displayed in Figure 4. The figure shows the reconstructed transverse, sagittal and coronal plane at the point of maximum dose for the two beams compared with their corresponding TPS dose distributions.

4. Discussion and Conclusion

3D imaging of dose inside a cylindrical water phantom was shown for a 3 x 3 cm^2 and complex beam apertures typically used in stereotactic radiosurgery plans. 3D Gamma (2%/2mm) passing rates exceeded 96% for all the reconstructed beams. Slice gamma analysis (2%/2mm) for the transverse, sagittal and coronal planes exhibited >94% passing rate. In future versions, a phantom jig would be constructed which would enable to phantom to be clamped to the edge of the couch, therefore enabling 360 ° imaging.

5. Disclosures

Brian Pogue and Petr Bruza report financial involvement with DoseOptics LLC, manufacturer of cameras and systems for Cherenkov imaging. Daniel Alexander and Cedar Farwell are paid consultants for DoseOptics LLC outside the scope of this work.

6. Acknowledgements

This work was funded by NIH research grant R01EB023909 and the Norris Cotton Cancer Center shared resource P30CA023108.

7. References

[1] Archambault L *et al* 2007 *Med. Phys.* **34** 1583-92
[2] Glaser AK *et al* 2013 *Med. Phys.* **40** 012101
[3] Beddar S *et al* 2009 *Med. Phys.* **36** 1736-43
[4] Glaser AK *et al* 2013 *Opt. Lett.* **38** 634
[5] Bruza P *et al* 2017 *IEEE Trans. Med. Imaging* **36** 2099-103

Development of an automated routine for the calibration of multi-point scintillation detectors for advanced dosimetry applications

B Lessard[1,2], F Larose[1,2], F Berthiaume[1,2,3], S Lambert-Girard[1,2,3], F Therriault-Proulx[3], L Archambault[1,2]
[1]Département de physique, génie physique et optique, et Centre de recherche sur le cancer, Université Laval
[2]CHU de Québec – Université Laval et CRCHU de Québec, Québec, Canada
[3]Medscint inc., Québec, Canada

Email: boby.lessard.1@ulaval.ca

Abstract. Multi-point scintillation dosimeters have the advantage to allow real-time measurements of dose with a good spatial resolution, but these detectors must be accurately calibrated to perform precise dose measurements. This study demonstrates, by means of simulations and an experimental validation, that when the calibration dataset is well chosen, it is possible to perform an automated calibration of such dosimeters, with an accuracy on the measured output factors reaching (0.5 ± 0.1) %.

1. Introduction

Scintillation dosimetry is taking advantage of the properties of plastic scintillators to convert ionizing radiation into visible light. This technology has the advantage to allow real-time measurements of dose with a good spatial resolution (i.e. submillimeter), due to the small size scintillators can have. In addition, plastic scintillation detectors also have the advantage of being water-equivalent (Beddar *et al.* 1992a, 1992b), which make them well suited for advanced dosimetry applications. To perform accurate dose measurements with such detectors, they must be precisely calibrated, however the calibration process can be long and tedious.

The goal of this study is to develop an automated calibration routine, using a translation module controlled by a computer, to enhance the calibration of multi-point scintillation dosimeters used with external radiation therapy photon beams. The module must allow the user to easily perform the required measurements for the calibration of the dosimeter when used under a photon beam delivered by a linac. Therefore, the calibration dataset conducting to the most accurate dose measurements with the dosimeter must first be determined.

2. Methods and materials

The principle of scintillation detectors is based on the collection of the light produced within a plastic scintillator using a waveguide; the collected light is thereafter measured by a photodetector. Plastic optical fiber is usually used as a waveguide to conduct the light from the scintillators to the photodetector without introducing non-water equivalent material. When more than one scintillating element is present, the detector is said to be "multi-point". Figure 1 shows an example of a multi-point scintillation detector (mPSD). Because the light produced by every scintillator travels the same unique waveguide, a hyperspectral approach has been proposed to separate the light coming from each scintillator

(Archambault *et al.* 2012, Therriault-Proulx *et al.* 2012). However, this method requires the pure spectra of each spectral component present within the total measured signal to be accurately known, which is a difficult task to perform with the high energy photon beam delivered by a linac because of the Cherenkov radiation light that is produced within the fiber. The method developed here aims at calibrating the mPSD without the prior knowledge of the pure spectra of the scintillators, using only a set of measurements that can be performed at the linac.

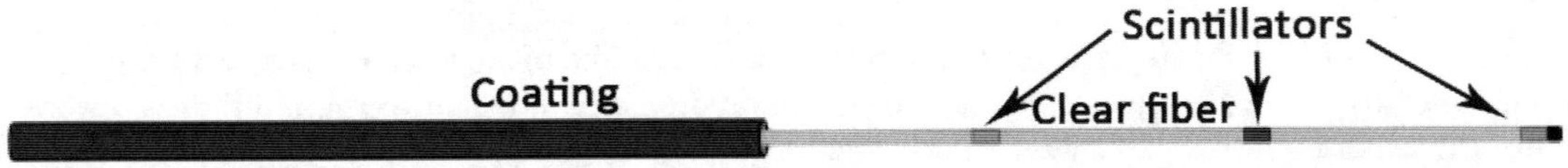

Figure 1. Representation of a typical multi-point detector. In that case, the detector contains 3 scintillators.

The detector used for this study is a 3-point scintillation detector, and the scintillators used are BCF-10, BCF-12 and BCF-60. In order to know what kind of measurements should be performed for the calibration of the mPSD at the linac, many calibration datasets have been simulated from the pure spectra of the scintillators and from the stem spectrum generated within the clear fiber. Measurements under a highly collimated kilovoltage photon beam at which there is no Cherenkov light emission produced within the optical fiber have been first acquired and are considered as the reference spectra for the scintillators, while the stem spectrum is measured at the linac using a subtraction method. Using a simple mPSD homemade simulator, coded in *Python*, allowing to simulate the measured dose under given experimental conditions with a calibrated mPSD, the calibration datasets have been classified according to their accuracy for subsequent dose measurements. For helping to choose the best calibration datasets for the mPSD, Principal Component Analysis (PCA) is performed on the data, using the pure spectra to build the Principal Components space. The transformed data is thereafter plotted in a simplex formed by the pure spectra composing the signal for a better visualization of the calibration datasets relative to the pure spectra.

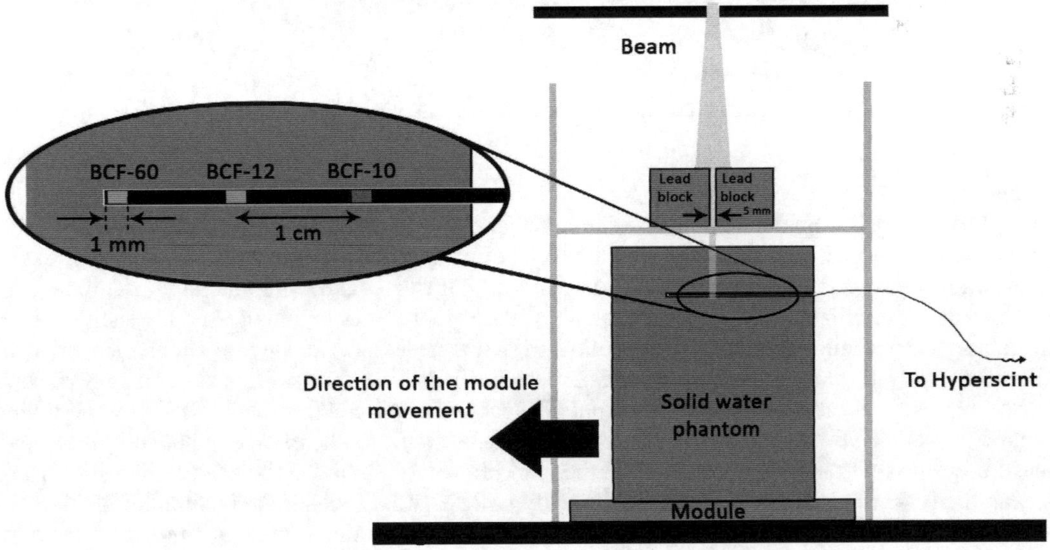

Figure 2. Schematic representation of the experimental setup.

Once the best calibration dataset has been determined by the simulations, using the representation of the data in the Principal Components space, a preliminary experimental validation is performed. The device used experimentally to measure the light produced by these scintillators is the *HYPERSCINT* scintillation dosimetry research platform (Medscint inc., Québec, Canada) allowing for both intensity and spectral measurements. An acquisition software linked to the translation module is used to measure the spectrum of the light coming from the fiber and to automatically control the translation module during the calibration. The translation module can therefore be programmed for performing the required irradiations for the calibration, in order to have an automated calibration routine. Lead blocks are also used into the setup to create a narrow slit of 5 mm into which the photon beam can pass through and get to the detector, while the detector, on the translation module, can move under this slit for the calibration. Once the calibration of the detector is performed, irradiations of the detector at different field sizes are performed in order to plot the output factors for the three scintillators and compare them to the reference. A schematic representation of the experimental setup is shown in figure 2, and figure 3 shows the real experimental setup, under the photon beam delivered by the linac.

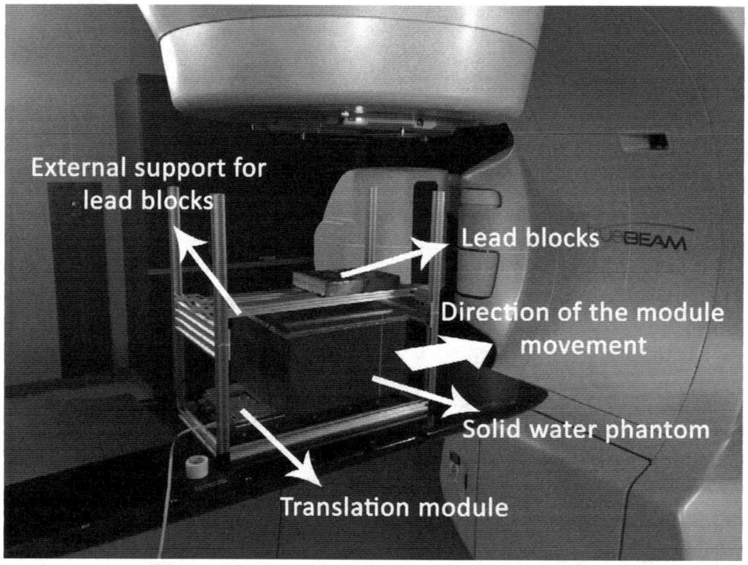

Figure 3. Experimental setup at the linac.

3. Results

The results obtained from the simulations showed that the best dosimetric results are obtained when the mPSD is calibrated with a dataset for which every points of measurement are situated near the corners of the simplex formed by the pure spectral components in the Principal Components space, which means that the scintillation light from each component of the signal has to be maximised for each measure of the dataset. The simulations also showed that it is not necessary to have a lot of measures in the calibration dataset, as long as the measures are well chosen. An example of a simulated calibration dataset, represented in the Principal Components space and which can be reproduced experimentally with a 3-points mPSD, is shown in figure 4 (a). Figure 4 (b) shows the experimental calibration dataset measured using the translation module and the lead blocks to maximise the scintillation light for each measurement, which is in good agreement visually with what has been done in simulations. Using this calibration dataset to calibrate the mPSD, the output factors obtained, both in simulations and at the linac, are shown in figure 5. The average error on the output factors, compared to the reference, is (0.5 ± 0.1)%, while this error was (0.09 ± 0.04)% in simulations.

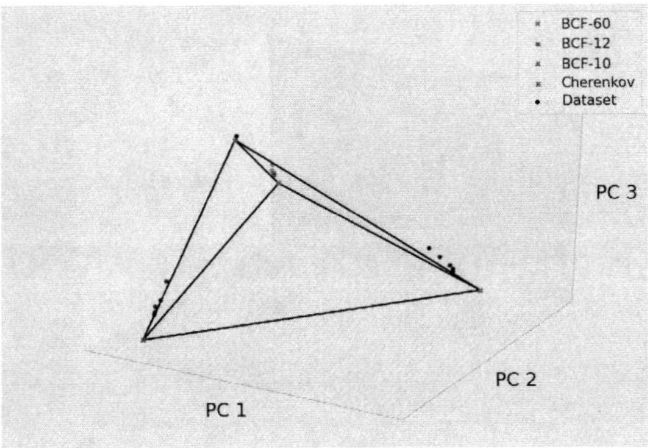

(a)

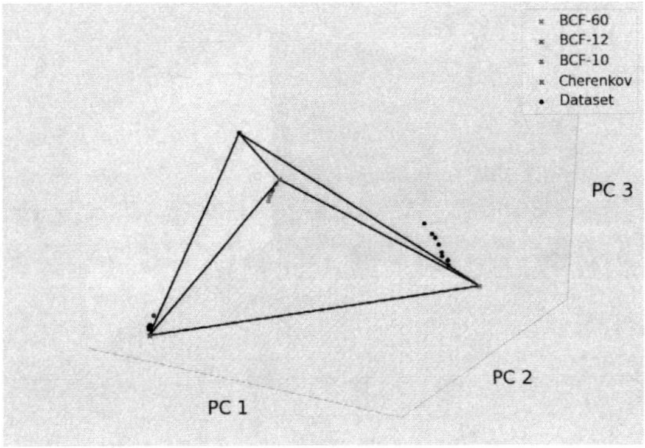

(b)

Figure 4. (a) Calibration dataset that gives the most precise dose measurements for the mPSD simulated with the homemade simulator. (b) Calibration dataset acquired at the linac, reproducing the best dataset obtained in simulations.

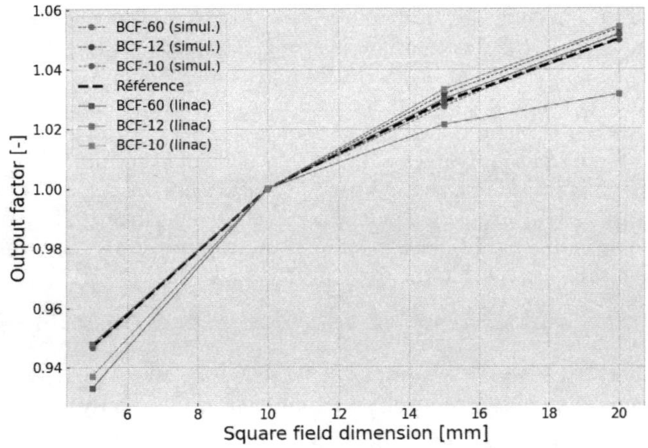

Figure 5. Output factor curves obtained with the three scintillators, at the linac and in simulations.

4. Conclusion

This study demonstrates that the translation module developed, along with its software, makes it possible to calibrate a multi-point scintillation detector with a high-energy photon beam. Furthermore, the use of the translation module facilitates greatly the calibration process, preventing the user from having to move the mPSD between each calibration measurement. While it is possible to make very small steps with the translation module (lesser than 1 mm if desired), it is not possible to move the jaws of the linac with steps as small as 1 mm, making the module and the lead blocks necessary for acquiring many measures around the scintillating elements of the mPSD in order to get calibration points near the corners of the simplex in the Principal Components space for the calibration. A further dosimetric validation could be done experimentally in order to verify the limits of the calibration of the mPSD with such a dataset for the calibration.

5. References

[1] Archambault L, Therriault-Proulx F, Beddar S *et al* 2012 *Med. Biol.* **57** 7133-46
[2] Beddar A S, Mackie T R and Attix F H 1992a *Phys. Med. Biol.* **37** 1883-900
[3] Beddar A S, Mackie T R and Attix F H 1992b *Phys. Med. Biol.* **37** 1901-13
[4] Therriault-Proulx F, Archambault L, Beaulieu L *et al* 2012 *Phys. Med. Biol.* **57** 7147-59

Journal of Physics: Conference Series **2167** (2022) 012029 doi:10.1088/1742-6596/2167/1/012029

Real time 4D Radiation Gel Dosimetry on the Australian MRI-Linac

Yves De Deene and Morgan Wheatley
School of Engineering, Macquarie University, North Ryde, Sydney, Australia

E-mail: Yves.DeDeene@mq.edu.au

Abstract. 4D radiation dosimetry using a highly radiation-sensitive polymer gel dosimeter with real-time quantitative MRI readout is presented as a technique to acquire the accumulated radiation dose distribution during image guided radiotherapy (IGRT) on an MRI-Linac. Optimized T_2 weighted TSE scans are converted into quantitative ΔR_2 maps and subsequently to radiation dose maps. The potential of real-time 4D radiation dosimetry in a theragnostic MRI-Linac is demonstrated in test tubes, for a square beam in a cylindrical gel phantom, for a simple step-and-shoot irradiation in a head phantom and a dynamic arc treatment on a cylindrical gel phantom using a rotating couch. The optimal sequence parameters for maximal dose resolution in the dynamic MRI acquisition will be presented and the trade off between MRI scanning speed and dose resolution will be discussed. A further improvement in temporal resolution using a keyhole imaging approach is the focus of future research.

1. Introduction

For decades, gel dosimeters have been proposed and applied in radiotherapy (RT) to display the radiation dose distribution in three dimensions [1, 2]. We here present the first results of online MRI of a radiation sensitive polymer gel dosimeter, imaged on the Australian MRI-Linac during radiation. While gel dosimetry has been applied on an MRI-Linac previously [3-5], this is the first study that demonstrates the feasibility of real-time 4D gel dosimetry. The possibility to image the gel dosimeter in pseudo-real time also sheds some light on short term chemical processes inside the polymer gel dosimeter. These early results demonstrate the potential of real time radiation dosimetry of MRI guided radiotherapy. An increase in readout speed is feasible with keyhole MRI.

In order to acquire real time dose maps, the dosimeter needs to have a high MRI dose sensitivity, exhibit a fast radiation response and be tissue equivalent. The post-irradiation response should be as low as possible. Moreover, like other 3D gel dosimeters, favorable characteristics are a good temporal stability, a high spatial integrity, a low temperature sensitivity and low dose rate dependence. We found that a Methacrylic Acid based polymer gel dosimeter [6] has a high dose-R_2 sensitivity and satisfies most of the aforementioned criteria.

2. Methods and Materials

2.1. Manufacturing of the Polymer Gel Dosimeter

Polymer gel dosimeters (PGDs) based on 6% (w/w) methacrylic acid and 8% (w/w) gelatin, referred to as MAGAT, were fabricated according to a recipe discussed elsewhere [6]. Four different sets of phantoms were fabricated: 14 test tubes that served as calibration vials, two 1L cylindrical flasks (diam. 10 cm) and a 3L head phantom.

2.2. Rotating Couch for the MRI-Linac

In the Australian MRI-Linac, the Linac (Linatron-MP, Varian Medical Systems) is static with respect to the patient table and MRI scanner. To enable a rotational dynamic treatment, an automated rotating phantom holder was constructed in house. The construction of the rotating phantom system will be discussed elsewhere [7].

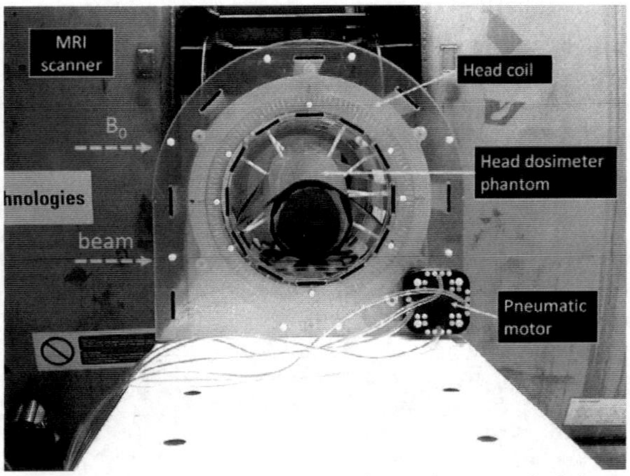

Figure 1. In house constructed rotating coach driven by a pneumatic motor.

2.3. MRI Scanning and Irradiation

MRI scanning and radiation of the PGDs was performed on the 1T MRI-Linac. The phantoms were also scanned 26 hours post-radiation on a clinical 3T MRI (Siemens Verio).

The *calibration vials* were irradiated in a standard water phantom at reference depth (10 cm). The dose rate at reference depth was 1 Gy/min. During radiation, the calibration gel dosimeters were scanned repeatedly inside the receive head coil with a T_{2w} Turbo-Spin Echo (TSE) sequence (TE: 271 ms; TR: 2 s; NEX: 2; Turbo Factor (TF): 22; Bandwidth per pixel (BW): 275 Hz/pixel: FOV: 200 mm; MS: 256×256; d: 5 mm). A single scan took 48 s. Immediately after radiation exposure, the gels were scanned with a Multi-Spin-Echo (MSE) sequence with 32 echoes spaced at 15 ms from which a calibration curve was obtained that was used to convert all dynamic scans to dose maps (Figure 2).

A 1L *cylindrical gel phantom* (diam. 120 cm) was irradiated with a 3 cm × 3 cm 6 MV photon beam in the longitudinal direction. The gel phantom was scanned dynamically with a TSE sequence with similar sequence parameters as above but with a FOV of 171 mm.

Another 1L *cylindrical gel phantom* (diam. 110 cm) was inserted in the rotating coach and irradiated with a lateral 2.6 cm × 2.6 cm beam that was placed 3 cm off axis with respect to the cylindrical phantom. The phantom was rotated in incremental steps of 9 degrees every 18 seconds while the beam was kept on during the entire radiation. The dose rate at isocenter for the radiation beam was set at 100 cGy/min. The total radiation treatment was split in 3 parts (each corresponding with a rotation over 120 degrees), between which the Linac was allowed to cool down for a few minutes.

The *head phantom* was exposed with 3 coplanar beams at angular increments of 120 degrees. For each of the 3 directions, the phantom was automatically rotated by use of the rotating couch. Dynamic TSE scans were taken with imaging parameters: FOV/MS/d/TR/TE/TF/BW = 220 mm/192/10 mm/1 s/160 ms/13/130 Hz. The time between two consecutive scans was 11 seconds with 10 seconds effective scanning and 1 second delay time.

The TSE images were converted to dose maps by a procedure as outlined in Figure 2. The calibration phantoms were also scanned with a multi-spin echo (MSE) sequence with 32 echoes and echo time spacing of 15 ms. R_2 maps were constructed from the 32 contrast weighted base images by use of an exponential fit on a pixel-by-pixel basis. The R_2 values for each of the calibration vials were extracted from circular regions of interest taken inside the test tubes and an R_2-dose plot was obtained. The change in R2 as a result of radiation can be extracted from the ratio of the signal intensity in a dynamic TSE scan recorded before radiation ($S(0)$) and the signal intensity ($S(t)$) in a dynamic TSE scan recorded during radiation at time t by use of equation 1.

$$\Delta R_2 = \frac{1}{TE}\ln\left(\frac{S(0)}{S(t)}\right) \tag{1}$$

A bi-exponential fit on the R_2-dose calibration plot was used to convert the ΔR_2 values to dose values.

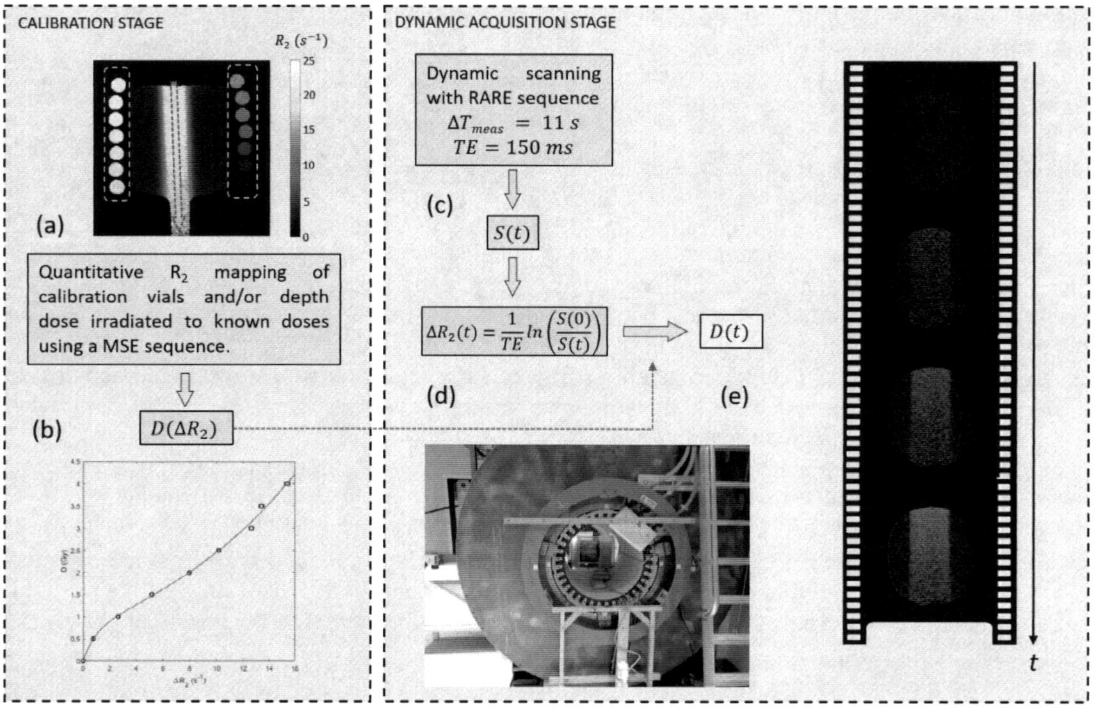

Figure 2. Procedure for acquiring real time dose maps on the MRI-Linac. First, calibration vials are scanned with a multi-spin echo sequence (a) from which a dose-R_2 calibration plot can be derived (b). Alternatively, a cylindrical flask can be irradiated with a single beam of which the dose distribution is known. During the treatment, dynamic scanning is performed with a TSE sequence which results in a series of T_2 weighted images (c) that can be converted to dose maps using the earlier derived calibration plot (d-e).

3. Results and Discussion

The dose-R_2 response of the MAGAT PGD follows a bi-exponential function on both the 1T MRI-Linac and on the 3T MRI scanner. The sensitivity of the MAGAT PGD in the linear region of the dose-R_2 response curve was found to be 3.95 $s^{-1}.Gy^{-1}$ on the MRI-Linac immediately after radiation and was found to be 4.5 $s^{-1}.Gy^{-1}$ 26 hours after irradiation on the 3T MRI scanner. The difference in dose sensitivity is attributed to post-irradiation polymerization. Real-time recorded dose maps show a nearly linear increase over time in recorded dose during radiation (Figure 3), and an exponential saturating response with a time constant of 80 minutes indicating post-irradiation polymerization. A model is derived to describe and fit the radiation response in the gel as a function of time during radiation (figure 3c). Assuming a first order kinetic response on radiation, it can be shown [8] that the registered dose is given by

$$D_r(t) = \dot{D}t + \dot{D}\tau_r\left(e^{-t/\tau_r} - 1\right) \qquad \text{for } 0 \le t \le t_{ON} \tag{2}$$

$$D_r(t) = \dot{D}t_{ON} - \dot{D}\tau_r e^{-t/\tau_r}\left(e^{t_{ON}/\tau_r} - 1\right) \qquad \text{for } t \ge t_{ON} \tag{3}$$

where $\dot{D}$ is the dose rate, τ_r is the time constant of the post-irradiation response and t_{ON} is the time that the beam is on.

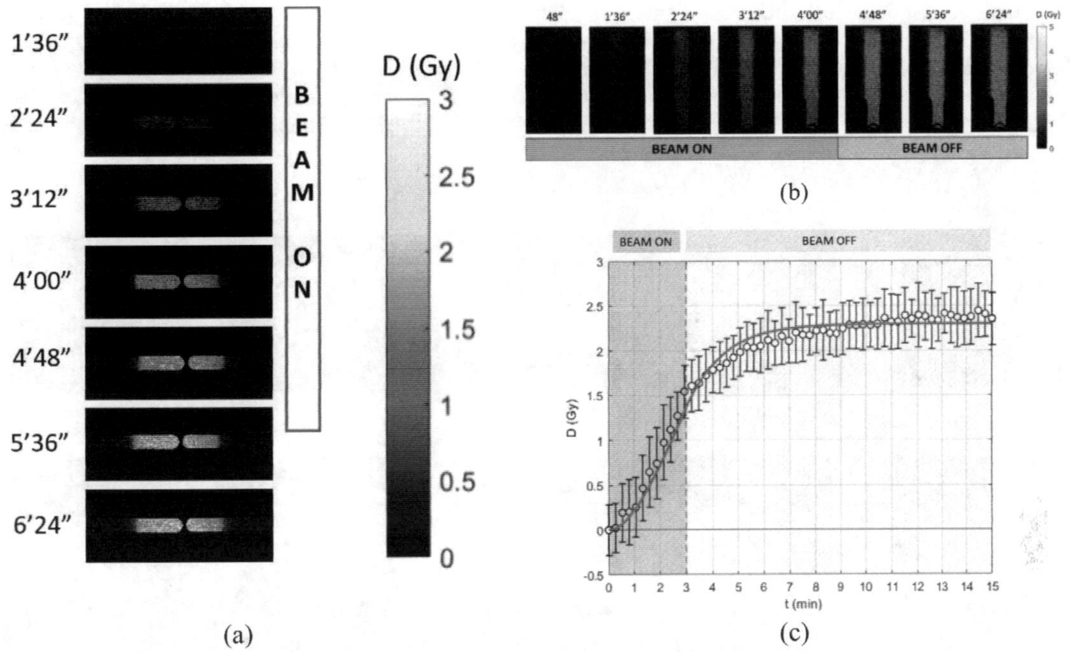

Figure 3. Sagittal measured dose maps obtained from T_{2w} TSE scans acquired at the 3T MRI-Linac. Calibration vials (a) were irradiated at reference depth in a water phantom. A cylindrical flask was irradiated with a small 2 cm × 2 cm beam (b). The change in registered dose over time demonstrates a post-irradiation response with a time constant of 83 s (c). The red shaded area is the response when the beam was on, while the green shaded area is when the beam was switched off.

Dose maps recorded during a dynamic arc treatment with off centre beam are shown in figure 4.

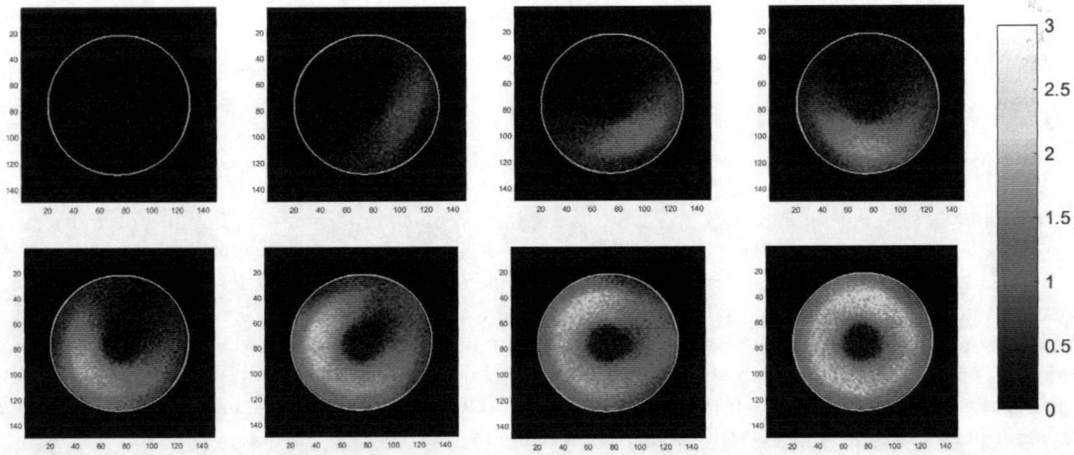

Figure 4. MRI measured dose maps (central slice) during a rotational treatment on a cylindrical phantom. The separation between two adjacent images is 7 recorded frames, corresponding with an angular increment of 54 degrees.

A movie of the dose formation can be found here:

https://www.dropbox.com/s/xe30c3ty63aezwc/Cylinder.avi?dl=0. The angular variation in dose in the final dose maps is attributed to an instability in dose rate during radiation which was confirmed from the logs on the Linac.

Dynamic dose maps of the head phantom at different time points during radiation are shown in figure 5. Every row corresponds to a different angle of radiation. The time between different rotation angles was approximately 14 minutes during which the beam was switched off. A movie containing all frames recorded at 11 second intervals is provided as supplementary material (https://www.dropbox.com/s/xwfxb98kgfltv6d/Head.avi?dl=0). The post-irradiation polymerization is visible between frames when the beam direction was changed (i.e. between the last image in a row and the first image of the next row in figure 5).

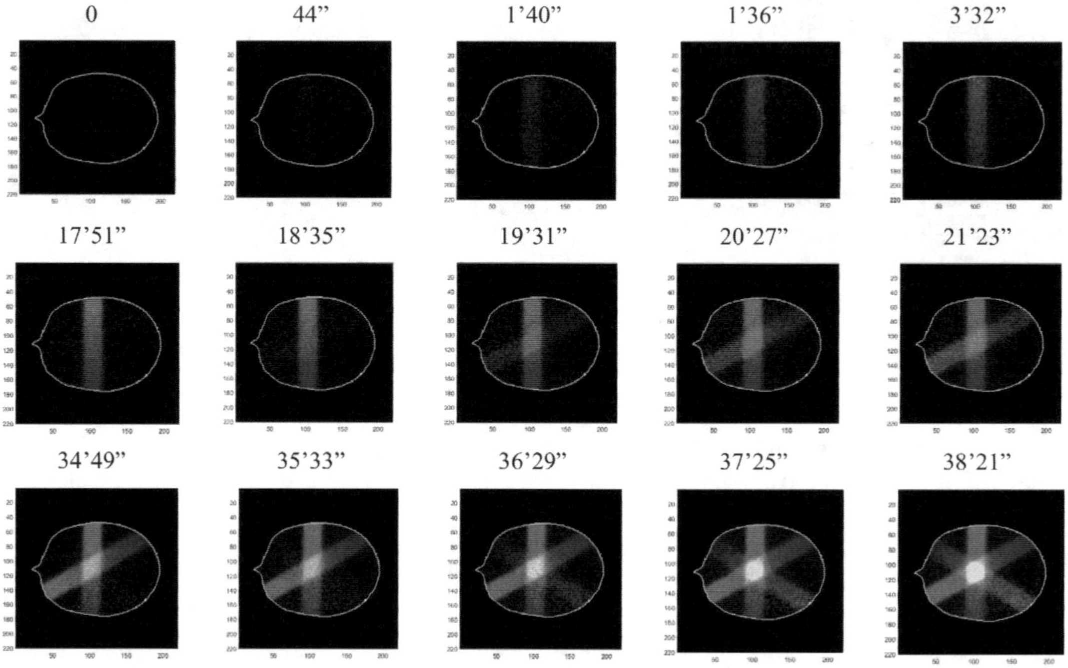

Figure 5. MRI measured dose maps recorded during radiation of a head phantom with three beams. Between each row, the phantom was rotated by 120 degrees to deliver the next beam. The time stamps above each dose map indicate the time since the start of radiation. No radiation was delivered between each rotation (i.e. between each row).

4. Conclusions

It is demonstrated for the first time that 4D radiation dosimetry with polymer gel dosimeters is feasible. The polymer gel dosimeter has a good stability but still shows some post-irradiation polymerization after radiation exposure. The time constant of this post-irradiation polymerization is in the order of 80 seconds. While these preliminary results are obtained on an experimental 1T MRI-Linac with relatively low dose rate, it can be expected that higher SNR is achievable at a 1.5 T MRI-Linac. A more extensive discussion on the important properties, optimization and limitations of gel dosimetry for real time dosimetry on an MRI-Linac can be found elsewhere [8]. For higher dose rates and for the compensation of organ motion, a higher frame rate without sacrificing SNR is achievable by using a keyhole approach. Such an imaging sequence is currently under development.

5. Acknowledgements

The authors thank Gary Liney, Bin Dong and Ursula Jelen from the Ingham institute for providing access to the Australian MRI-Linac. All consumables for this research were funded from the first authors’

Journal of Physics: Conference Series **2167** (2022) 012029 doi:10.1088/1742-6596/2167/1/012029

salary as a result of a lack of Australian research funding. The first author thanks his family for supporting this research financially.

6. References

[1] De Deene Y, De Wagter C, Van Duyse B *et al* 2000 *Magn. Reson. Med.* **43** 116-25
[2] Baldock C, De Deene Y, Doran S *et al* 2010 Polymer gel dosimetry *Phys. Med. Biol.* **55** R1-63
[3] Lee HJ, Kadbi M, Bosco G *et al* 2018 *Phys. Med. Biol.* **63** 045021
[4] Ibbott GS *et al* 2019 *J. Phys. Conf. Ser.* **1305** 012011
[5] Dorsch S, Mann P, Elter A *et al* 2018 *Phys. Med. Biol.* **63** 045021
[6] Dorsch S, Mann P, Elter A *et al* 2019 *J. Phys. Conf. Ser.* **1305** 012007
[7] De Deene Y, Vergote K, Claeys C *et al* 2006 *Phys. Med. Biol.* **51** 653-73
[8] Wheatley M 2020 *J. Phys. Conf. Ser.* (In press)
[9] De Deene Y, Wheatley M, Liney G *et al* 2020 *Phys. Med. Biol.* **65** 225031

Development of photonic detector system for ultra-fast beam diagnostics in proton radiotherapy: the proof of concept

Viktor Iakovenko[1] and David A. Jaffray[2]
[1]Department of Radiation Oncology, University of Toronto, ON, M5T 1P5, Canada
[2]Radiation Physics and Imaging Physics, UT MD Anderson Cancer, Houston, TX, USA

E-mail: vic.yakovenko@gmail.com

Abstract. A concept of a photonic detector system for proton beam and Bragg peak position measurements in proton radiation therapy is presented. An approach of using scintillator plates with ultra-fast timing characteristics to detect the temporal fine structure of the beam is described. A detector module is made of a 10 × 10 cm^2 plastic scintillator plate with 1mm thickness. The light is collected on the corners of a plate by the optical fibers of pre-defined length, which introduce various known time delays. Using the Anger algorithm, the lateral position of the proton pencil beam traversing scintillator plate is reconstructed. We propose two applications of the system: thin single-plate beam position monitor and multi-plate stack quality control device to measure lateral beam position and relative position of the Bragg peak.

1. Introduction

Radiation therapy remains an important component of cancer treatment with approximately 50% of all cancer patients receiving radiation therapy during their course of illness [1]. Proton therapy is rapidly growing modality that is used to treat certain types of cancers [2]. As of May 2021 there are 98 operational proton therapy centers, 83 of those became clinical in the last 10 years [3]. Key advantages of proton therapy, relative to conventional photon radiotherapy, is the specific dose distribution though the patient with low entrance dose and with majority of the dose deposited at the end of the proton track – at the depth of Bragg peak [4]. The commissioning and operation of these facilities require an extensive set of detectors for measuring various parameters of the treatment beam, in particular beam position monitors and detectors for beam range quality control are among key devices [2]. Given the complexity of proton therapy, a comprehensive quality assurance program is required, to maintain the best quality of treatment delivery for cancer patients.

Due to large dose gradients in the region of Bragg peak, its precise location is of critical importance [2]. Historically, reference percentage depth doses for proton pencil beam are measured with ionization chamber in the water at different depths and transverse positions [5]. Scanning Bragg curve with this method is time consuming, therefore many proton centers utilize Multi-Layer Ionization Chambers (MLIC), which provide considerably faster method for beam range measurements [6]. However, MLICs have some disadvantages: ionization chambers are not water-equivalent, finite size of charge-collecting electrode can complicate the measurement of integrated depth-dose curves [7] and system's response exhibit dose-rate effects.

In the last decade, there has been an increase interest in scintillator-based detectors for quality assurance applications [2, 8-15], which can provide a competitive alternative to MLICs. A scintillator is a material that emits light in response to the deposition of energy by ionizing radiation. Modern scintillator detectors have a combination of features advantageous for proton beam measurements, including excellent timing characteristics, high spatial resolution, radiation hardness, high degree of water-equivalence and can be produced in various geometries and dimensions [16, 17]. It was shown by Beddar *et al.* [18] that plastic scintillators exhibit high sensitivity, linear dose-response and excellent energy independence compared to other dosimeters used in radiotherapy. Main disadvantage of scintillators in their application in proton dosimetry is light quenching effects, which leads to a non-linear dependence between light output and dose deposition for heavy charged particles, such as protons [19].

Solid plastic scintillators were used to verify proton beam range [2, 20], and liquid-scintillator detectors were used to measure simultaneously and, almost in the real time, the range and lateral profiles of proton beams [21].

In this paper we present a concept of 2D and 3D monitoring systems for ultra-fast beam diagnostics that leverage the temporal fine (bunch-per-bunch) structure of the proton beam. Components of the systems were optimized to meet specific timing requirements and performance of the systems was evaluated in the Matlab simulations. Application of the 2D system focuses on beam position monitoring of the proton pencil beam; 3D system – on the range measurements (similar to MLICs) with simultaneous lateral beam position measurements. Scintillation light from each plate is collected at the corners by the optical fibers of various pre-defined length, creating known optical signal delays and consequently readout by a single photomultiplier tube. This configuration avoids optical artifacts, since fibers are directly coupled to the scintillator plate. Signal from the light pulse train is reconstructed to determine the lateral position of the beam in a single plate. Light quenching is accounted using method described by Tretyak [22]. Integrated photonic signals in each plate are associated with the total energy deposited by the proton beam and used to determine the position of Bragg peak.

2. Materials and Methods

There are two basic factors that need to be taken into consideration when using plastic scintillators for timing measurements: the total number of photons generated in the scintillator, i.e. light yield; and the de-excitation time of the scintillator molecules that defines how fast photons are emitted, i.e. rising time.

2.1 Scintillator

Five types of scintillators suitable for fast timing measurements were considered: Detec PS-91F [23] and Saint Gobain Crystals BC-422Q (quenched 0.5%), BC-418, BC-420, BC-422 [24]. Physical properties of these scintillator materials are presented in Table 1. Birks' constant (kB = 9 mg/MeV/cm^2) to correct for quenching effect was obtained using semi-empirical method described by Tretyak [22].

Table 1. Scintillators available for ultra-fast timing measurements.

Scintillator	Rise time, τ_r [ps]	Decay time, τ_d [ns]	Light output, L_{yield} [% anthracene]	Attenuation length (bulk), [cm]	λ_{MAX} emission, [nm]
PS-91F	700	1.8	45	----	390
BC-422Q (0.5%)	110	0.7	19	≤ 8	385
BC-418	500	1.4	67	100	391
BC-420	500	1.5	64	110	391
BC-422	350	1.6	55	8	370

For the purpose of simulations performed in this work, scintillator with the fastest rising time (110 ps) – BC-422Q (0.5%) was selected.

2.2 Timing requirements
We have analysed the beam micro-structures of the TRIUMF cyclotron and synchrotron at Heidelberg Heavy-Ion Therapy Centre to estimate the scope of the timing requirements for proposed photonic detectors. Fine structure of the beam is defined by the accelerator's radiofrequency system (RF) that operates at about 23 MHz for TRIUMF cyclotron and 5 MHz for HIT synchrotron. This dictates the requirement for using scintillator with sub-nanosecond rising time and for system's readout timing performance from 40 ns to 200 ns, necessary to prevent overlapping of the photonic signal trains between the neighbouring bunches.

2.3 Photomultiplier tube (PMT), digitizer and optical fibers
The most important features for the timing measurements with PMT are rise-time, transit time spread (TTS), quantum efficiency (Q.E.), photocathode size and gain. A lower TTS and higher photocathode efficiency are essential for the timing measurements. Three typical PMTs used for ultra-fast timing measurements were considered with their characteristics listed in Table 2.

Table 2. Characteristics of PMTs suitable for ultra-fast timing measurements.

Hamamatsu PMT	Rise time, [ps]	TTS, [ps]	Q.E., [%]	Gain	Photocathode size, [mm]	Peak sensititvity, λ_{MAX}, [nm]
R7400U-04 [25]	780	230	26	5x10^5	8	400
R7400U-01 [25]	780	230	26	5x10^5	8	400
R3809U-52 [26]	150	25	22	2x10^5	11	385

To fulfill our timing requirement, we selected MCP-PMT R389U-52 with the rise time 150 ps, TTS-25ps and gain 2x10^5.

For the combination of BC-422Q (0.5%) scintillator and MCP-PMT R389U-52 the readout system should be capable of collecting the data with at least 5 GS/s sample rate. To satisfy this requirement, in our simulation we have selected a high-speed digitizer from National Instruments - PXIe-5186 with the sample rate 12.5 GS/s and 8 bit resolution.

The light from the corners of scintillator plate to PMT is delivered by multimode quartz optical fiber (FiberTech® UV-SIV) that is designed to transmit UV light of 350 – 390 nm. Attenuation length of the fiber A = 0.05 dB/m and aperture 0.22.

The photonic signal is delayed in the optical fiber. Some value of fiber lengths and introduce optical delays are presented on Table 3.

Table 3. Length of the fibers selected for every corner (Figure 1) and related time delays.

Corner #	1	2	3	4
Length (m)	0.1	1.6	3.6	6.1
Time delays (ns)	0.5	8.4	18.9	32.1

Ultrashort pulses propagating though the fiber can undergo substantial temporal and spectral changes, mostly due to chromatic dispersion. In high-bit-rate systems chromatic dispersion has a significant influence on the broadening of the pulse. Taking into account the dispersion of scintillator emission spectra $\sigma_\lambda = 60$ nm and the sample rate at 5GS/s, we have estimated the broadening of 97 ps per 1 m of the optical fiber.

3. Results and Discussion
The light emitted from the scintillator volume is collected on the corners of the scintillator plate, transmitted by optical fibers and readout with a single PMT (Figure 1). The collected light carries information about the beam's lateral position and energy. We propose creating optical delays with

Journal of Physics: Conference Series **2167** (2022) 012030 doi:10.1088/1742-6596/2167/1/012030

optical fibers of pre-defined lengths to distinguish corners of the plate and use that information to reconstruct the position of a proton beam from the photonic signal pulse train.

3.1.1. Single-Plate Design

We designed the detector in a form of a square scintillator plate with dimensions 10×10 cm^2 and thickness 0.1 cm (Figure 1). Traversing the plate, proton beam creates a photonic signal in the scintillator (Figure 2). The light from the interaction point *beam-plate* travels to the plate corners, which are coupled to the optical fibers. In Matlab we have simulated a transport of the photonic signal train as it is formed and delayed by the optical fibers of pre-defined length. Only prompt photons from the interaction point that are in the acceptance angle of each corner are detected. The length of the fibers was optimized to be sufficient to contain and transmit the total photonic signal generated by a single proton bunch before the arrival of the consecutive bunch, i.e. within one RF period (200 ns for HIT synchrotron). Since the length of each fiber, attenuation length and signal behavior inside fiber, as well as amount of light lost due to quenching are known, it is possible to reconstruct the true signal generated in the scintillator plate at every corner. Total amount of light produced per bunch and temporal structure of the signal is used to evaluate the energy of the impacting proton beam. To define the position of interaction point, an Anger algorithm was utilized [27]. Proton beam interaction point was deliberately simulated off plate's centre (Figure 2) to study the capabilities of the system to identify off-axis lateral beam position.

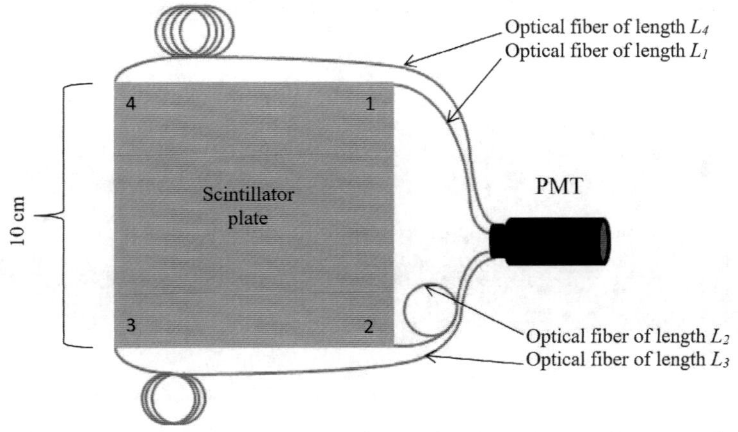

Figure 1. Design of the beam position monitor. Scintillator plate is readout on the corners by optical fibers of different length, herewith $L_1 < L_2 < L_3 < L_4$. Bunch per bunch photonic signal arrives to the PMT in a form of signal pulse train.

The standard Anger algorithm can be expressed by the following equation:

$$x = \frac{(N_1+N_2)-(N_3+N_4)}{N_1+N_2+N_3+N_4}, y = \frac{(N_1+N_4)-(N_2+N_3)}{N_1+N_2+N_3+N_4}, \tag{1}$$

where N_1, N_2, N_3, N_4 are the reconstructed true number of photons at corners 1, 2, 3 and 4 respectively (Figure 1). x, y are the coordinates of the proton beam entering the plate (Figure 2). For the scintillation light originating directly from the interaction point in rotated by $\pi/4$ plane, the position x^* and y^* can be defined using a positioning algorithm described by Zhang *et al.* [28], which also reduces the 'pin-cushion' distortion. The new algorithm can be expressed as an asymmetry functions in (x^*, y^*) coordinates:

$$x^* = \frac{N_1-N_3}{N_1+N_3}, y^* = \frac{N_2-N_4}{N_2+N_4}, \tag{2}$$

Original x, y coordinates can then be calculated as:

$$x = x^* \cos\left(\frac{\pi}{4}\right) + y^* \sin\left(\frac{\pi}{4}\right), y = y^* cos\left(\frac{\pi}{4}\right) + x^* sin\left(\frac{\pi}{4}\right). \quad (3)$$

Resolution of the system to identify the centroid of the proton beam lateral position is determined by two factors: variation in the amount of scintillation light entering optical fibers as proton beam traverses the plate off-centre and resulting in solid angle variation at each corner of the plate; and shift of photonic pulses on the temporal scale of uncentered beam relative to its central temporal signal response. Our simulations suggest that utilizing of these two mechanisms < 5 mm resolution can be achieved.

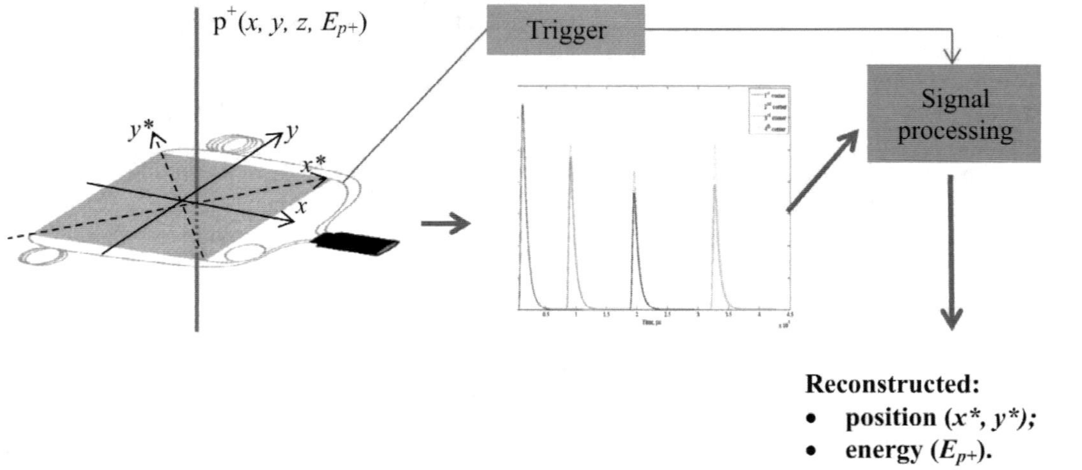

Figure 2. Schematic signal flow in the single-plate detector. Proton beam transverses scintillator plate off-centre producing light. Photonic signal from the shortest optical fiber is used to trigger the system. Photonic signal from the plate corners is used to reconstruct interaction point position (x^*, y^*) and energy deposited in the plate by the proton bunch, which can be associated with proton beam energy.

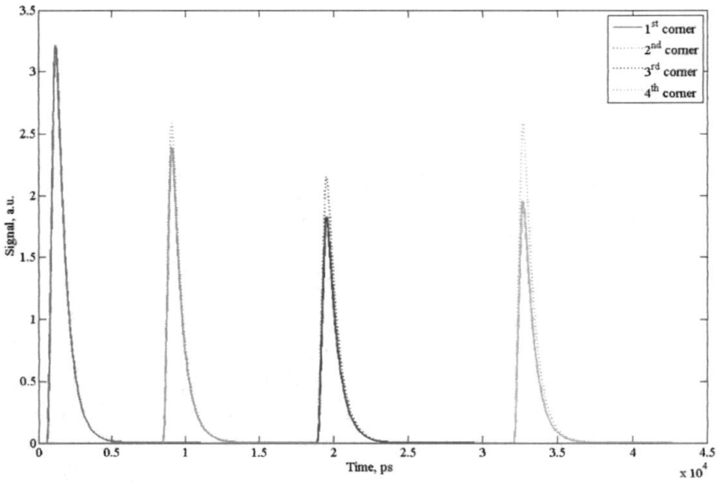

Figure 3. Simulated photonic signal train of an uncentered proton beam. Thick line represents the corrected for quenching signal registered by the PMT, while dotted line – simulated signal corrected for quenching in scintillator plate and signal modifications during the transport in the optical fibers.

Time delay (τ_d), created by the longest fiber guiding light from the scintillator plate corner to the PMT plus the bunch length (t_b) should be less than the time between bunches (T): $\tau_d + t_b < T$. This requirement prevents overlapping of the signals from the neighboring bunches. By the time the signal from the next coming bunch arrives through the shortest fiber to the PMT, all light generated by the first bunch in all fibers would already be detected.

3.1.2. Multi-Plate Design

An array of a single-plate detector modules stacked together in the direction of the proton beam can extend the application of the system by enabling beam range measurement (Figure 4). In our design multi-layer structure is immersed in water-filed rectangular cuboid vessel with the length sufficient to cover the range of proton beam. Longitudinal resolution of detector system depends on the thickness of scintillator plate and longitudinal spatial granularity of plates in the beam's direction. Scintillator plate thickness of each module is 1 mm, while granularity can vary depending on requirements: it can be sparse in the plateau region of Bragg curve or can be stacked back-to-back as implemented by Kelleter *et al.* [2]. In the latter, optical decoupling between modules should be maintained. The position of the plate with maximum integral photonic signal, i.e. maximum amount of energy deposited in the plate, can be associated with the Bragg peak position with 1 mm accuracy. Calibrated system with the known proton energy can be utilized for daily quality control.

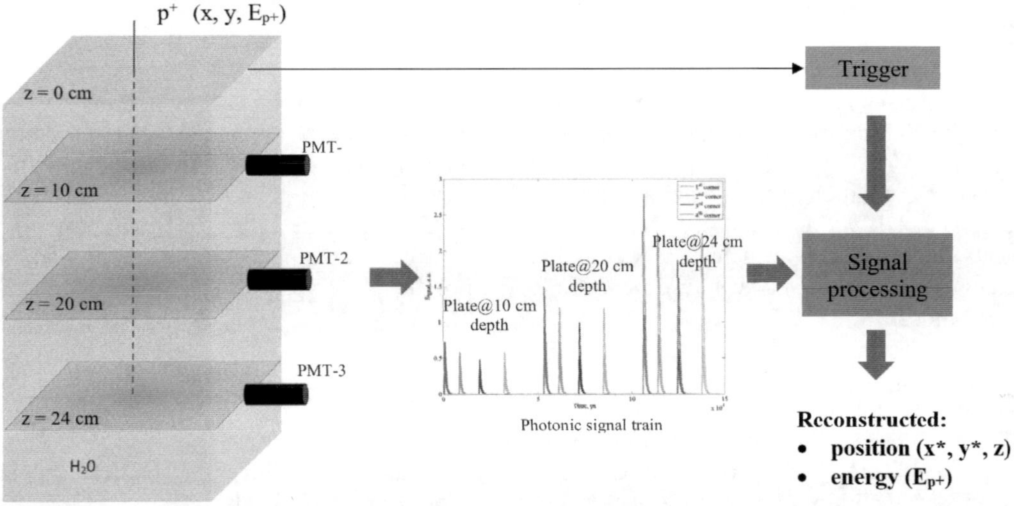

Figure 4. Schematic signal flow in the three-plate detector system. 200 MeV proton beam traversing the water phantom deposits increasing amount of energy with increasing depth. System readout is triggered by every proton bunch crossing entrance phantom window at $z = 0$ cm. Each module is readout by a single PMT.

We have simulated a water phantom with the three scintillator plates at depths of 10 cm, 20 cm and 24 cm in the 30 cm length water phantom (Figure 4). Our simulations suggest that system can readout every 150 ns. In Figure 5, we have simulated the response of a 28-plate system in 200 MeV proton beam with modules positioned at 1 cm increments. Transformation of photonic signal in the optical fibers is known from the section 2.1.3 and can be utilized to reconstruct the original signal in each scintillator plate and reconstruct the Bragg curve, position of Bragg peak with 1 mm resolution and lateral position of the beam with 5 mm resolution.

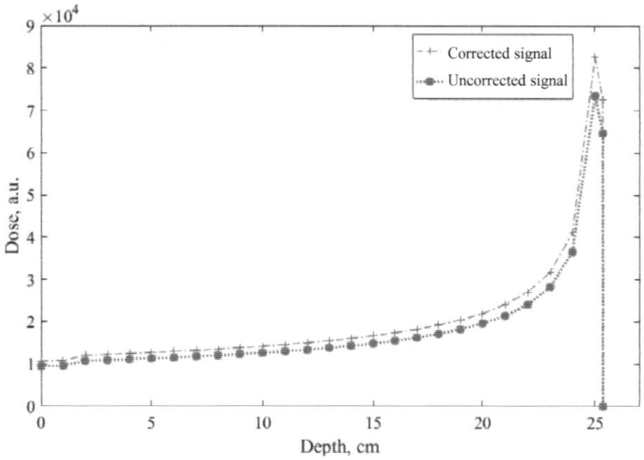

Figure 5. Bragg curve from a 28-layer detector system simulated for 200 MeV proton beam. In red – uncorrected (except for quenching) simulated integral PMT signal, blue – simulated integral signal corrected for quenching in scintillator plates and for attenuation, broadening in the optical fibers, representing true photonic signal emitted from the scintillator plate.

Proposed system is flexible, scalable and provides opportunity to quantitatively examine typically unknown fine structure of the particle beam. In addition, the system do not exhibit dose-rate effects, which is an important feature for FLASH proton therapy when action may be required on the nanosecond temporal scale to provide safe and effective treatment.

4. Conclusion

The results of our simulation suggest the possibility of using a proton fine beam structure for diagnostics of proton beam. Single-plate design of proposed detector system can be applied as a beam monitor for high intensity proton pencil beams, capable to measure simultaneously lateral position of the beam's lateral centroid position with < 5 mm accuracy and energy of the beam by utilizing temporal structure of the light signal. Multi-plate design can be applied as a quality control device for Bragg peak position verification with ~1 mm accuracy defined by spatial frequency of the plates in the system and by the scintillator plate thickness. Our future steps are focused on the prototyping of these devices to validate our simulations.

5. References

[1] Baskar Rajamanickam *et al* 2012 *Int. J. Med. Sciences* **9** 193-9
[2] Kelleter L *et al* 2020 *Phys. Med. Biol.* **65** 165001
[3] Particle Therapy Co-Operative Group: Particle therapy facilities in clinical operation www.ptcog.ch/index.php/facilities-in-operation. Assessed on May 27, 2021.
[4] Kristensen I, Nilsson K, Nilsson P 2015 *Int. J. Part. Ther.* **2** 367-75
[5] Karger C P, J¨akel O, Palmans H *et al* 2010 *Phys. Med. Biol.* **55** R193-234
[6] Actis O, Meer D, Koenig S *et al* 2017 *Phys. Med. Biol.* **62** 1661-75
[7] Baeumer C, Koska B, Lambert J *et al* 2015 *J. Appl. Clin. Med. Phys.* **16** 151-63
[8] Beaulieu L, Beddar S 2016 *Phys. Med. Biol.* **61** R305-43
[9] Almurayshid M 2017 *J. Appl. Clin. Med. Phys.* **18** 210-19
[10] Alsanea F 2018 *Med. Phys.* **45** 1782-9
[11] Henry T 2017 *Int. J. Part. Ther.* **4** 1-6
[12] Guillot M, Beaulieu L, Archambault L *et al* 2011 *Med. Phys.* **38** 6763-74
[13] Lacroix F, Sam Beddar A, Guillot M *et al* 2009 *Med. Phys.* **36** 5214
[14] Robertson D, Mirkovic D, Sahoo N *et al* 2013 *Phys. Med. Biol.* 5 261-73
[15] Therriault-Proulx F, Archambault L, Beaulieu L *et al* 2012 *Phys. Med. Biol.* **57** 7147
[16] Beddar S, Archambault L, Sahoo N *et al* 2009 *Med. Phys.* **36** 1736-43
[17] Beddar S 2015 *J. Phys. Conf. Ser.* **573** 012005
[18] Beddar AS, Mackie TR, Attix FH 1992 *Phys. Med. Biol.* **37** 1883-900

Journal of Physics: Conference Series **2167** (2022) 012030 doi:10.1088/1742-6596/2167/1/012030

[19] Birks JB 1951 *Proc. Phys. Soc. A.* **64** 874
[20] Fukushima Y, Hamada M, Nishio T *et al* 2006 *Phys. Med. Biol.* **51** 5927-36
[21] Archambault L, Poenisch F, Sahoo N *et al* 2012 *Med. Phys.* **39** 1239-46
[22] Tretyak VI 2010 *Astroparticle Physics* **33** 40-53
[23] DETEC Scintillation *Materials Brochure.* Retrieved from Detec Web site, 2006.
[24] Saint-Gobain Crystals Scintillation *Materials and Assemblies Brochure.* Retrieved from Saint-Gobain Crystals Web site, 2008.
[25] Hamamatsu Photonics *Metal package photomiltiplier tube R7400U datasheet.* Retrieved from Datasheet catalog Web site, 2001.
[26] Hamamatsu Photonics *Metal package photomiltiplier tube R3809U-52 datasheet.* Retrieved from Datasheet catalog Web site, 2015.
[27] Anger HO 1964 *J. Nucl. Med.* **65** 515-31
[28] Zhang J, Olcott PD, Levin CS 2007*IEEE Trans. Nucl. Sci.* **54** DOI: 10.1109/tns.2007.894129

Journal of Physics: Conference Series **2167** (2022) 012031 doi:10.1088/1742-6596/2167/1/012031

New directions for tetrazolium - gellan gum gel dosimeters

Kalin I. Penev[1,2] and Kibret Mequanint[1,3]
[1]Department of Chemical and Biochemical Engineering, the University of Western Ontario, London, ON, N5A 5B9, Canada
[2]Modus Medical Devices Inc., 1570 N Routledge Park, London, ON, N6H 5L6, Canada
[3]Biomedical Engineering Program, the University of Western Ontario, London, Ontario, N6A 5B9, Canada

E-mail: kmequani@uwo.ca

Abstract. Tetrazolium salt-based dosimeters are non-diffusing, gel dosimeters with excellent physical and chemical stability but with relatively low dose sensitivity and a dose rate dependence of the dose response. Both issues are tackled in this study by: (a) the introduction of a new tetrazolium salt with simple synthesis and dose response around 585 nm which may provide the pathway to more sensitive formulations and (b) producing gels with low gellan gum concentration, thus limiting the dose rate effects, and addition of thickening agents to restrain the liquid loss (syneresis) from the resulting gels.

1. Introduction

Tetrazolium salts (TS) are attractive as radiochromic sensor compounds owing to their low background coloration and diffusion-free gel dosimetry, as in gellan gum gel with "bisnitrotetrazolium chloride" (BNC) [1,2]. A version of this composition with 0.25 mM BNC has been commercialized as ClearView 3D Dosimeter by Modus QA (London, Canada) and has found application for the dosimetry of high-dose stereotactic treatments, e.g. [3]. More recently, we tested a series of tetrazolium salts containing strong chromophores [4]; however, as research in chemical dosimeters shows, hundreds of related compounds may need to be tested before the most favorable radiochromic chemistry is found [5]. With that in mind, we performed quantum chemical modelling of a series of candidate compounds [6], which revealed that changes at the group bound to the carbon atom of the tetrazolium ring have minimal effect on the electrostatic potential (and therefore, sensitivity) of the tetrazolium salt, but may either facilitate or hinder its synthesis and can shift the spectrum of the resulting formazan. Electron-donating groups, such as 4-methoxyphenyl, may be particularly useful, as they would allow for easy synthesis and a bathochromic spectral shift in absorbance (to higher wavelengths). In the current study we have compared that previously used BNC with its 5-(4-methoxyphenyl) analogue (Figure 1). We have also recently demonstrated that the sensitivity of BNC-gellan gum gels can be increased by addition of methylparaben to act as a radical scavenger, while the total solids loading can by cross-linking with lithium salts without compromising with optical clarity [7]. The resulting gels, while mechanically strong, show a tendency to release liquid (syneresis) when subjected to external force. Here, we test whether the addition of thickening agents can alleviate the syneresis without affecting the dose rate dependence or any other properties of the dosimeter. In short, the goal of this study is to present the preliminary investigation a new and promising tetrazolium salt and approaches to limit the dose rate sensitivity by decreasing the gellan gum concentration and to minimize syneresis by introduction of a thickening agent.

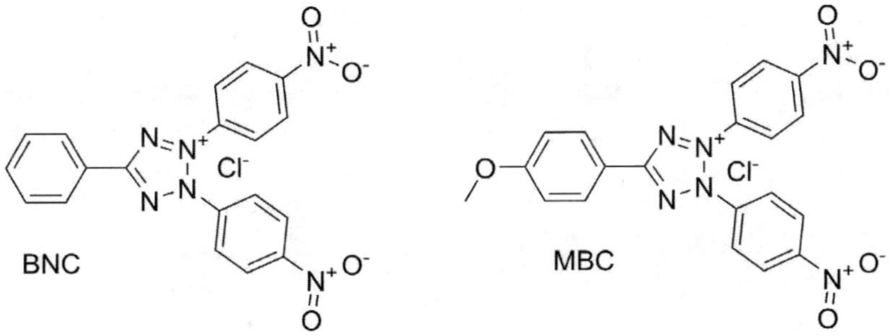

Figure 1. Tested compounds: 2,3-bis(4-nitrophenyl)-5-phenyl-2*H*-tetrazolium chloride (BNC) and 5-(4-methoyxphenyl)-2,3-bis(4-nitrophenyl)-2*H*-tetrazolium chloride (MBC)

2. Materials and Methods

MBC was synthesized according to a previously published procedure using 4-methoxybezladehyde, 4-nitrophenylhydrazine and 4-nitroaniline according to [8], whereas all other materials were purchased from commercial suppliers and used without further modification: BNC as "2,3-Bis(4-nitrophenyl)-5-phenyltetrazolium chloride hydrate" (cat. no. B1047) from TCI America, gellan gum (cat. no. J63423) and methylparaben (cat. no. A14289) from Alfa Aesar, propylene glycol (cat. no. P355) from Fisher Scientific, and polyacrylamide (150 kDa, cat. no. 767379) from Millipore Sigma. The tested compositions are given in Table 1. Gel preparation, cuvette specification, storage conditions, optical readout and irradiation setup were as described previously [1].

Table 1. Tested compositions.

	Gellan gum, wt.%	Propylene glycol, wt.%	Methylparabe n, wt.%	Polyacryl amide, wt.%	[LiCl], mM	TS	[TS], mM
A1	1.25	10	0	0	0	BNC	0.25
A2	1.00	10	0.25	0	25	BNC	0.25
A3	0.75	10	0.25	0	50	BNC	0.25
A4	0.75	10	0.25	0.25	50	BNC	0.25
B1	1.25	10	0	0	0	MBC	0.25
B2	1.00	10	0.25	0	25	MBC	0.25
B3	0.75	10	0.25	0	50	MBC	0.25
B4	0.75	10	0.25	0.25	50	MBC	0.25

As a predictor for syneresis, the liquid expulsion during centrifugation was measured under using Amicon Ultra-4 centrifugal filter units (10 kDa pore size, 3.0 cm^2 filtration area, cat no UFC8010) from Millipore Sigma filled with *ca.* 1.6 g of tested gel and centrifuged at 2000×*g*. For statistical purposes, the effect of the TS type was ignored in the study of syneresis: two samples from each **A** and one from each **B** composition were tested, providing sets of triplicates.

3. Results and Discussion

Figure 2 shows the background optical density, aging at room temperature, and dose response of compositions **A1** and **B1**. The wavelength of maximum absorbance shifts from *ca.* 530 to *ca.* 585 nm, allowing the gels to be scanned with amber light and decreasing the optical background and scatter within the gel matrix. The rate of auto-reduction was independent of the composition and varied among samples at $(5.1 \pm 1.2) \times 10^{-4}$ cm^{-1} day^{-1} in units of optical density at the respective wavelengths of interest. Whereas the sensitivity of MBC gels appears lower, it should be noted that the material can be prepared readily and in high yields unlike BNC, which means that MBC may be used at an increased concentration in order to obtain more sensitive gels at little added cost. Figure 3 shows that decreasing the gellan gum concentration leads to lower dose rate dependence, while the presence of polyacrylamide has no apparent effect on the sensitivity.

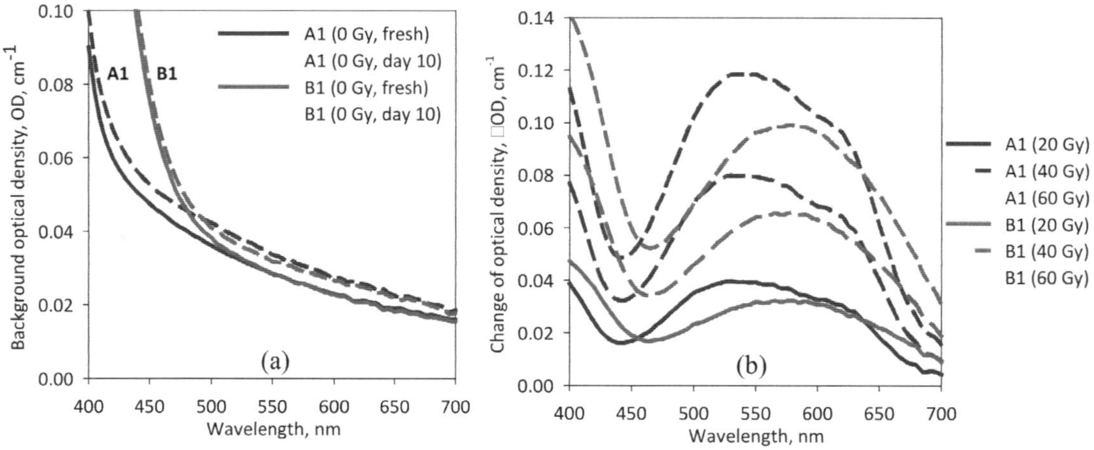

Figure 2. Comparison of BNC and MBC dosimeters (**A1** and **B1**): optical density and auto-reduction in panel (a) and dose response in panel (b). Irradiations on a Cobalt-60 source at *ca.* 100 cGy/min.

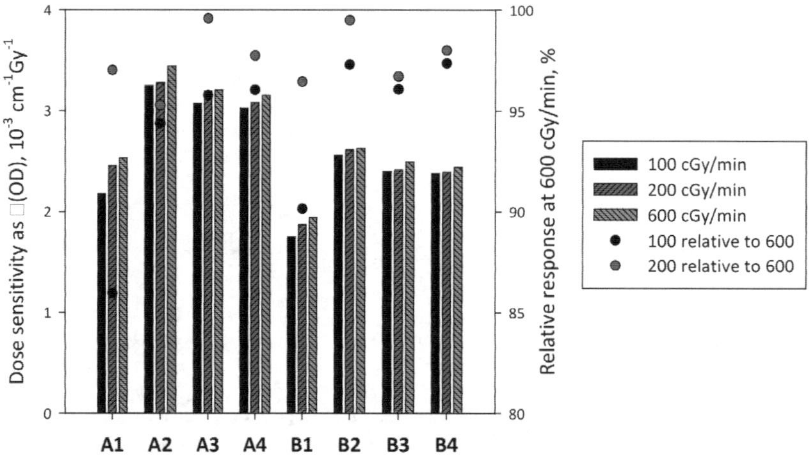

Figure 3. Dose rate effect of the dose response. Irradiation was performed using 6 MV x-ray to cuvettes positioned 5.0 cm under water, source-to-surface distance was 95.0 cm to the water surface.

Figure 4 supports the potential utility of compositions **A4** and **B4**, as the addition of polyacrylamide improved the resistance of the gel to syneresis, as modelled according to the equation:

$$w_{loss} = \frac{w_{eq}}{\tau + t_c} t_c \qquad (1)$$

Where, w_{loss} is the mass loss, w_{eq} is the equilibrium mass loss, τ is a characteristic time period, t_c is the cumulative centrifugation time, and w_{eq}/τ is the initial rate of mass loss. The results are presented in

Figure 4 and Table 2. Comparing gels **A3/B3** and **A4/B4** shows that in the absence of polyacrylamide the gels cannot reach equilibrium water content at 0.75% gellan gum concentration. More importantly, the initial rate of mass loss is drastically decreased between the two compositions, indicating resistance to syneresis in composition A4/B4. It should be noted that in addition to polyacrylamide other thickening agents were tested with the goal of decreasing the rate of syneresis. For example, carboxymethylcellulose and Pluronic F-127 showed promising results but both were ultimately rejected due to decreased optical quality of the gel in the first case, and reaction with the tetrazolium salts in the second.

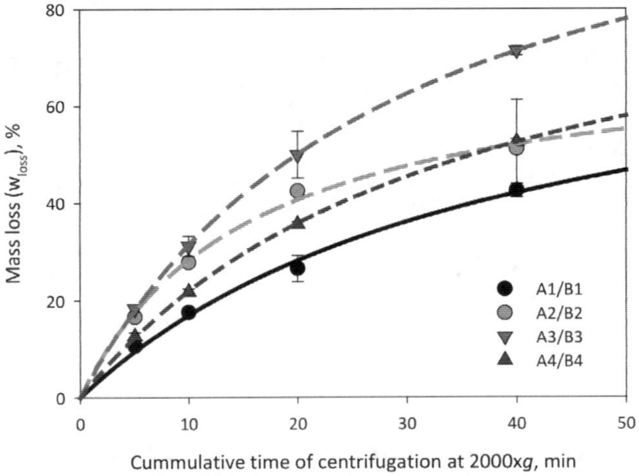

Figure 4. Results from the syneresis study.

Table2. Modelling results.

	w_{eq}, %	w_{eq}/τ, %/min	R^2
A1/B1	63±8	2.5±0.4	0.996
A2/B2	79±5	4.2±0.1	0.998
A3/B3	117±5	4.3±0.1	1.00
A4/B4	92±5	2.9±0.0	1.00

4. Conclusions

The use of polyacrylamide as a thickening agent in gellan gum gels presents a promising perspective for decreasing the total solids loading in the dosimeters, and indirectly lowering the dose rate effects. Further, the while MBC showed lower sensitivity than BNC, the bathochromic shift in absorbance and ease of preparation make this compound attractive for further studies.

Based on the promising results from the cuvette experiments, larger scale, complex dose-delivery plans will be tested using gellan gum – polyacrylamide gels at low total solids loading to prove the promising dose-rate effects, with both tetrazolium salts presented in this study.

5. Acknowledgements

We would like to acknowledge the help of Mr. Khalid Noori in performing experimental tests and data analysis for this work. Funding for this research was provided by the Collaborative Health Research Projects (CHRP).

6. References

[1] Penev K I, Wang M, Mequanint K 2017 *J. Phys. Conf. Ser.* **847** 012048
[2] Ascención Y, Dietrich J, Mequanint K *et al* 2017 *J. Phys. Conf. Ser.* **847** 012049
[3] Du D, Quinn B, Penev K *et al* 2019 *Med. Phys.* **46** E310
[4] Hazarika R, Penev K I, Mequanint K *et al* 2019 *J. Phys. Conf. Ser.* **1305** 012036
[5] Adamovics J A, Coakley R J 2019 *J. Phys. Conf. Ser.* **1305** 012028
[6] Penev K I, Bedada T and Mequanint K 2020 *Chem. Phys.* **535** 110790
[7] Brzozowski P, Penev K I, Mequanint K 2021 *Int. J. Biol. Macromol.* (under review)
[8] Penev K I, Mequanint K 2015 *J. Het. Chem.* **53** 1655

Journal of Physics: Conference Series **2167** (2022) 012032 doi:10.1088/1742-6596/2167/1/012032

RSC: Dosimetry in high-dose-rate brachytherapy with a radio-fluorogenic gel dosimeter

K. Inoue[1], Y. Watanabe[1,2], T. Maeyama[3], S. Mizukami[2], S. Hayashi[4], T. Terazaki[5], H. Muraishi[1,2], T. Gomi[1,2] and T. Shimono[1,2]
[1]Graduate School of Medical Sciences, Kitasato University, Kanagawa, Japan
[2]School of Allied Health Sciences, Kitasato University, Kanagawa, Japan
[3]School of Science, Kitasato University, Kanagawa, Japan
[4]Faculty of Health Sciences, Hiroshima International University, Hiroshima, Japan
[5]Yokohama City Minato Red Cross Hospital, Kanagawa, Japan

E-mail: ar16812@st.kitasato-u.ac.jp

Abstract. A nanoclay-based radio-fluorogenic gel (NC-RFG) was used to verify the source position and dose distribution in high-dose-rate (HDR) brachytherapy. The dose response confirmed linearity up to 60 Gy. The source position could be detected with an accuracy of ≤0.3 mm, and the dose distribution near the Ir-192 source showed good agreement with the Monte Carlo simulation. NC-RFG can be expected to be a quality assurance tool suitable for the evaluating the dose distribution in HDR brachytherapy.

1. Introduction

In high-dose-rate (HDR) brachytherapy, a sealed radioisotope (the source) is transported directly inside or near a tumor using a remote after-loading system (RALS). Its dose distribution shows steep dose gradients, depending on the distance from the source. The source is exchanged every 3-4 months, because the half-life of iridium-192 (Ir-192) is 73.8 days. Source positioning errors lead to severe consequences, including a reduced therapeutic effect due to an insufficient dose or the occurrence of complications resulting from the delivery of a high dose to normal tissues [1,2]. Therefore, it is advisable to verify that there is no abnormality in the radiation emitted from the new Ir-192 source at each exchange, and a quality assurance program for the equipment used in treatment is required for the correct execution of HDR brachytherapy [3-5].

A dosimeter with a wide dynamic range is necessary for evaluating the high-dose area near the vicinity of the source. At present, the detailed measurement of the dose distribution near the source is not possible. Thus, the HDR brachytherapy dose distribution calculation algorithm is used, which is based on the principle of superposition of dose distributions near the source according to the original and updated recommendations of Task Group 43 (TG-43 and TG-43U1) of the American Association of Physicists in Medicine (AAPM) [6,7].

In recent years, the application of radio-fluorogenic gel dosimeters as new dose distribution measurement tools has been expected [8-13]. The purpose of this study was to verify the dose distribution and source position of Ir-192 using a nanoclay-based radio-fluorogenic gel (NC-RFG), which is a fluorescent gel dosimeter using dihydrorhodamine 123 hydrochloride (DHR 123) as a fluorescent probe [13,14].

2. Material and methods

2.1. Gel preparation
This study used the NC-RFG proposed by Maeyama et al. [13]. The NC-RFG was prepared using 2.5% w/w nanoclay (Laponite XLG; Rockwood Ltd., UK), 97.5% w/w ultra-pure water (Purelab Flex UV, Elga Lab., UK) and 0.1 mM DHR 123 (FUJIFILM Wako Pure Chemical, Japan). The non-fluorescent DHR 123 is changed to fluorescent rhodamine 123 (RD 123) by radiation-induced oxidation. The NC-RFG used in the experiment was enclosed in the 3 mm gap of a custom-made polymethylmethacrylate (PMMA) square layer container (7 × 100 × 100 mm) (Fig. 1).

2.2. Irradiation and data scanning
The RALS (microSelectron HDR V3; Elekta Brachy, Veenendaal, The Netherlands) and Ir-192 source were used for all experiments in this study. The dwell time and absorbed dose were calculated by a treatment planning system (Oncentra, Elekta Brachy, Veenendaal, The Netherlands). The Ir-192 source activity was 36.075 $mGy{\cdot}m^{2}{\cdot}h^{-1}$. One flexible catheter (ProGuide Sharp Needle; Elekta Brachy, Veenendaal, The Netherlands; length, 200 mm; diameter, 2 mm) was inserted to transport the Ir-192 source into the NC-RFG PMMA container (Fig. 1a). The source was transported to a single point in the center of the PMMA container for the acquisition of the dose response. The integrated irradiation doses were 5, 10, 20, 30, 40, 50, 60, 80 and 100 Gy for one sample at 10 mm from the center of the source. The irradiated NC-RFG was scanned with a GELSCAN-3 (iMeasure, Japan). GELSCAN-3 scans were performed at 465 nm excitation with a blue light-emitting diode (LED). The 48-bit RGB images scanned with GELSCAN-3 were 4 inches × 6 inches (2400 pixels× 3600 pixels), and the resolution was 0.04 mm/pixel. The green pixel components of RGB images of the NC-RFG were separated using the ImageJ software program (National Institutes of Health, USA).

2.3. Verification of dose distribution and source position
Fluorescence intensity was converted to dose using a response curve. The dose distribution near the source was measured using a single source position. The measured values of NC-RFG were compared with the calculated values of a Monte Carlo (MC) code (PHITS version 3.020) [15]. The MC simulation generated 10^8 photons irradiating a geometry shown in Fig. 1a with NC-RFG as water equivalent. The statistical uncertainty was 4.6% [13]. The dose normalized at 90 degrees for a transport direction of 10 mm from the source center was evaluated at 10-degree intervals. The source was transported at 2, 3, 4, 5 points at 40-, 20-, 10- or 20-, and 10-mm intervals in a range of 40 mm along a single catheter for verification of the source position (Fig. 1b). The fluorescence intensity peaks at source positions were detected by full width at half maximum (FWHM) using a 3-mm dose profile.

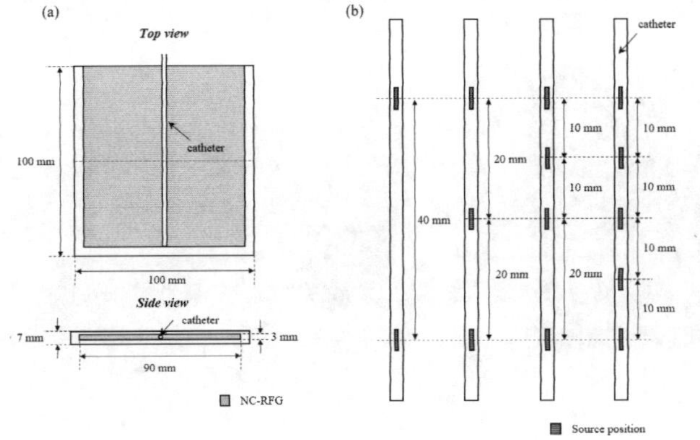

Figure 1. (a) A schematic diagram of the catheter setting in a nanoclay-based radio-fluorogenic gel dosimeter (NC-RFG) enclosed in an acrylic container. The green area is NC-RFG. (b) The Ir-192 source transportation plan in the range of 40 mm, for the verification of source positions.

3. Results and Discussion

3.1. Dose response

The fluorescence intensity according to the number of scans of the non-irradiated NC-RFG is shown in Fig. 2. Increased fluorescence intensity was confirmed by scanning without irradiation with Ir-192. It is possible that the amount of RD123 produced increased because the absorbance increased with the number of scans (the irradiation time of the excitation light). Samples scanned multiple times have problems with dose conversion. Thus, the fluorescence intensity of the NC-RFG sample irradiated with the Ir-192 source was evaluated by subtracting the increase in fluorescence intensity with respect to the number of scans (the increase was calculated by using the approximation formula in Fig. 2). From Fig. 3, the obtained fluorescence intensity and dose confirmed the linearity up to 60 Gy (correlation coefficient R^2 = 0.996). This result indicates that NC-RFG has a wide dynamic range and that it is suitable for measuring HDR brachytherapy dose distributions.

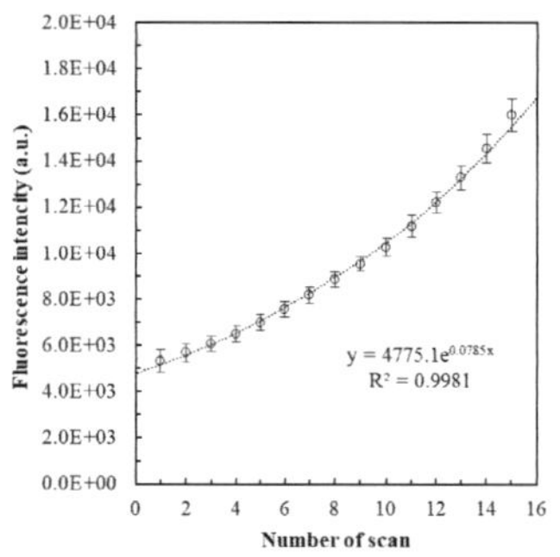

Figure 2. Increase in fluorescence intensity due to scanning. Fluorescence intensity of a non-irradiated sample for each scan.

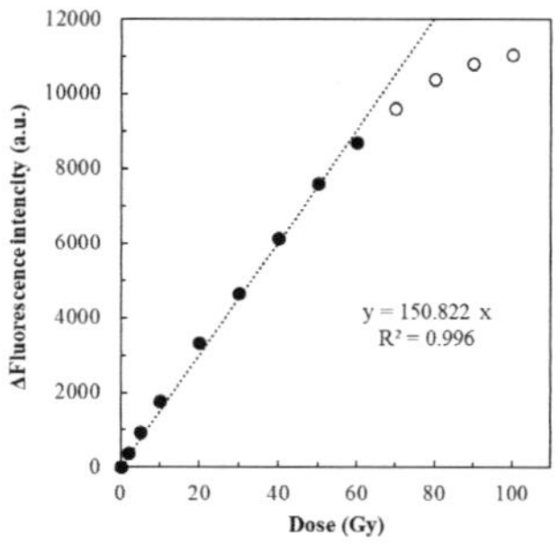

Figure 3. The dose–Fluorescence intensity response curve for the NC-RFG.

3.2. Dose distribution near the source

The Ir-192 source was 0.6 mm in diameter and 3.5 mm in length and encased in a stainless-steel capsule that was welded to a stainless-steel cable. Thus, gamma rays emitted from Ir-192 were emitted anisotropically. The calculated dose of MC and the fluorescence distribution of NC-RFG decreased in the longitudinal direction of the source (Fig. 4ab). The dose calculated by MC simulation and NC-RFG measurement were in good agreement between the two dose distributions around source (Fig. 4c). The possibility of using NC-RFG to evaluate the dose distribution around the source in detail was shown.

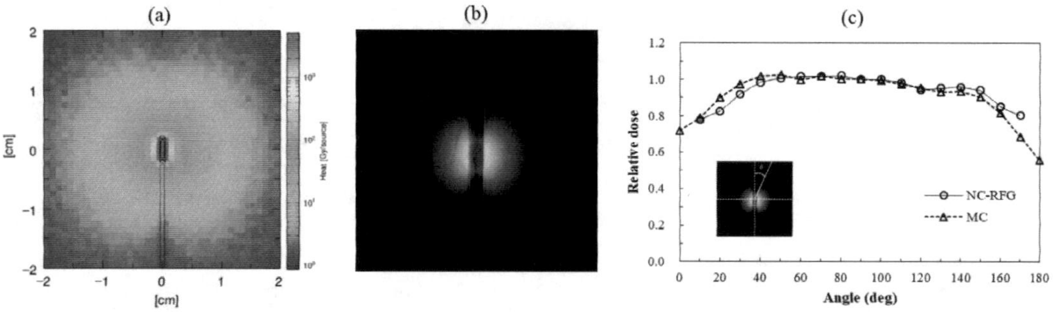

Figure 4. (a) MC calculation and (b) the fluorescence distribution of NC-RFG around a single Ir-192 source. (c) Comparison of the TPS calculations and NC-RFG measurements of gamma rays emitted anisotropically from the center of the Ir-192 source.

3.3. Verification of the Ir-192 source position

Fig. 5 shows fluorescence images of NC-RFG irradiated at multiple source positions. Clear fluorescence was confirmed according to the position of the source. The central part was not fluorescent due to the placement of the catheter. Each peak of the 3-mm dose profile at the source positions was confirmed in all samples. The comparison of the source position calculated from FWHM with the reference position set in RALS was 0.25 ± 0.08 at 2 points, 0.25 ± 0.09 at 3 points, 0.26 ± 0.23 at 4 points, and 0.26 ± 0.29 at 5 points. This is less than the 1-mm tolerance recommended for equipment quality assurance by TG-56 of the AAPM [4].

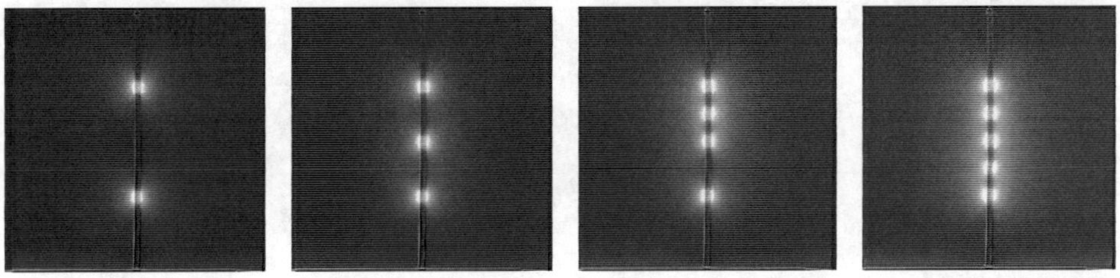

Figure 5. Fluorescence images of NC-RFG irradiated at multiple source positions.

4. Conclusion

In this study, NC-RFG was used to verify source location and dose distribution in HDR brachytherapy. NC-RFG has proven to be able to respond to the characteristics of HDR brachytherapy, which is a dose distribution with a high dose and a steep dose gradient.

5. Acknowledgments

This work was supported by JSPS KAKENHI [grant number JP18K07769 and JP17K07013]; and Kitasato University School of Allied Health Science [Grant-in Aid for Research Project number 2019-1045]. We wish to thank Dr. T. Furuta at Japan Atomic Energy Agency for his support on PHITS for the Monte Carlo calculations, and I. Shiga, K. Takeda, M. Mori, Y. Kasai and D. Shimizu at Kitasato Univ. for their support in performing the experiments.

6. References

[1] ICRP 2000 *Ann. ICRP* **30** 1-70
[2] ICRP 2005 *Ann. ICRP* **35** 1-51
[3] Kutcher G J *et al* 1994 *Med Phys.* **21** 581-618
[4] Nath R *et al* 1997 *Med Phys.* **24** 1557-98
[5] Venselaar J and Perez-Calatayud 2004 *ESTRO Booklet No. 8: A Practical Guide to Quality Control of Brachytherapy Equipment* (Brussels: ESTRO)
[6] Nath R *et al* 1995 *Med Phys.* **22** 209-34
[7] Rivard MJ *et al* 2004 *Med Phys.* **31** 633-74
[8] Warman JM *et al* 2011 *Phys Med Biol.* **56** 1487-508
[9] Warman JM *et al* 2009 *Phys Med Biol.* **54** 3185-200
[10] Warman JM *et al* 2019 *Polymers* **11** 1729
[11] Yao T 2017 *Delft University of Technology*
[12] Maeyama T *et al* 2018 *Radiat Phys Chem.* **151** 42-6
[13] Maeyama T *et al* 2019 *Sens Actuators A-Phys.* **298** 111435
[14] Watanabe Y *et al* 2020 *Phys Med Biol.* **65** 175008
[15] Sato T *et al* 2018 *J. Nucl. Sci. Technol.* **55** 684-690

Signal requirements for 3D optically stimulated luminescence dosimetry

Mads L. Jensen[1], Rosana M. Turtos[1], Jacob S. Nyemann[1], Brian Julsgaard[1,2], Ludvig P. Muren[3,4], Peter Balling[1,2]
[1]Department of Physics & Astronomy, Aarhus University, Denmark
[2]Interdiciplinary Nanoscience Center (iNANO), Aarhus University, Denmark
[3]Danish Centre of Particle Therapy, Aarhus University Hospital, Denmark
[4]Department of Medical Physics, Aarhus University/Aarhus University Hospital, Denmark

E-mail: mlj@phys.au.dk

Abstract. For 3D optically stimulated luminescence (OSL) based dosimeters to be clinically applicable, certain standards must be met. Among these are low detectable doses with high accuracy and precision, ideally comparable with those of point-like detectors. By investigating a model of the central part of an OSL readout-system, we present an estimate of $\sim 4 \cdot 10^7 \frac{\text{photons}}{\text{Gy}\cdot\text{mm}^3}$ as the minimum required signal from OSL active materials embedded in a transparent matrix to allow measuring doses of 0.1 Gy with an accuracy and precision of 2 %. Further, 2D spatially resolved measurements of OSL emission from commercially available LiF:Mg,Cu,P pellets are presented and discussed.

1. Introduction

Three dimensional dosimetry is a valuable tool in modern radiotherapy where treatment techniques are becoming increasingly complex. Quality assurance is key in such treatments and 3D dosimetry is an optimal way to provide this. Currently available 3D dosimeters consist of more or less tissue equivalent gels which change optical density according to the radiation dose [1]. The dose deposited in the gels can then be measured by the changes in optical density through optical-computed-tomography scanning. There are challenges of such gel dosimeters, and numerous studies on correcting these are available [1–4]. Unfortunately, one major drawback, namely the permanent nature of the changes in optical density, cannot be corrected. Solutions have been suggested [5,6], but the search for a reusable 3D dosimeter is ongoing.

As proposed by our group, optically-stimulated-luminescence (OSL) active micro/nanoparticles embedded in a transparent matrix could provide a solution to many of the challenges associated with this problem [7, 8]. Metastable trapped states in the active material will be populated when irradiated, and information about the deposited dose is stored in the density of these populated trapped states. To read out this information, the trapped states are emptied by optically stimulating recombination of the trapped electron-hole pairs. The intensity of light emitted from such recombinations will then hold information about the deposited dose. From this point on, this light will be referred to as the OSL signal. Optically manipulating a laser to form a light sheet using a Powell lens and collimating optics will make it possible to read out a single plane of the dosimeter at a time. Moving the dosimeter and hence emptying another plane, would then provide the third dimension. Such a dosimeter would be reusable, as it can be reset by bleaching with either light or heat.

In the following, the minimum OSL signal required from the active material for dosimetric applicability will be discussed. This is based on an analysis of the central part of the optical readout setup, i.e. light transmission from the dosimeter to the camera including the intermediate optical elements. Moreover, preliminary spatially resolved measurements of commercially available OSL active LiF:Mg,Cu,P (MCP) pellets will be presented and discussed.

2. Considerations on OSL signal

For a dosimeter to be clinically feasible for patient-specific quality assurance, it must be able to detect low doses with high accuracy and precision. In order to provide a lower bound on the dosimeter signal, this analysis is based on more point-like detector characteristics, requiring detectable doses as low as 0.1 Gy with an accuracy and precision of 2 % in a relevant voxel volume. This corresponds to a signal-to-noise-ratio (SNR) of 50, which under the assumption of a shot-noise limited readout-camera yields a minimum detected photoelectron number of $\mathrm{N} = \sqrt{\mathrm{SNR}} = 2500$. This number dictates the minimum detected signal needed from a voxel in the dosimeter, which can be converted to a minimal OSL signal needed from the active particles by backtracking through the system illustrated in Figure 1 accounting for signal losses in all elements.

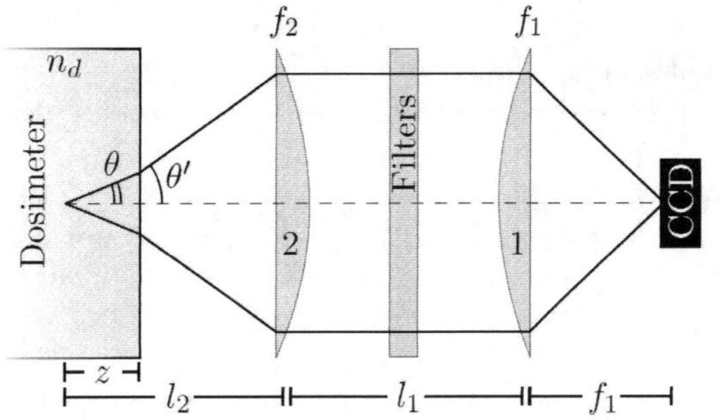

Figure 1. Sketch of the central part of the optical readout system illustrating the optical path from a point in the dosimeter to the CCD.

Starting from the dosimeter side, there will be some attenuation of the signal inside the dosimeter described by Lambert-Beer's law of attenuation. The transmitted signal will be given by the coefficient

$$k_{\mathrm{att}} = \exp(-\mu z), \qquad (1)$$

where z is the depth in the dosimeter and μ is the material-specific attenuation coefficient. This is under the assumption that the opening angle, θ_{max}, corresponding to a solid angle of the lens area, is small. Through a ray-matrix analysis, it can be shown that θ_{max} relates to the lens diameter d as $\theta_{\mathrm{max}} = \frac{d}{2n_d f_2}$, where n_{d} is the refractive index of the dosimeter and f_2 is the focal length of lens 2. This is valid under small-angle and thin-lens approximations. As the active material is emitting isotropically, the fraction of signal that will be detectable, i.e. emitted in the solid angle related to θ_{max}, is given by the coefficient

$$k_\Omega = \frac{1}{2}(1 - \cos\theta_{\mathrm{max}}) \approx \frac{1}{16} \frac{m^2}{n_{\mathrm{d}}^2 f_{\#1}^2}, \qquad (2)$$

where $m = -f_1/f_2$ is the magnification and $f_{\#1} = f_1/d$ is the f-number for lens 1.

Reflection at the interface between the dosimeter and air will also cause signal loss. Again, considering small angles, it can be justified to estimate this loss using the normal-incidence case of the Fresnel equations, yielding the following transmission coefficient

$$k_{\mathrm{T}} = \frac{4n_{\mathrm{d}}}{(n_{\mathrm{d}}+1)^2}. \qquad (3)$$

Both lenses, the filters, and the quantum efficiency of the camera will give rise to further signal losses and the transmission can be described by coefficients, k_l, k_f and k_{QE}, respectively, which can be measured or looked up in data sheets.

Regarding spatial resolution, one mm^3 is a clinically relevant voxel size. The required number of photons emitted from such a voxel per given Gy of dose can now be estimated as

$$N_{\text{voxel}} = \frac{2500}{0.1\ \text{Gy}} \prod_i \frac{1}{k_i}, \tag{4}$$

where i runs over all correction coefficients. A spatial resolution of one mm^3 is readily achievable with the transverse resolution ensured by a large CCD and matching optics, and the longitudinal given by light sheet thickness.

Using silicone (Sylgard 184) as a typical elastomer, the refractive index and attenuation coefficient is measured at a wavelength of 370 nm to be 1.45 and 0.07 cm^{-1}, respectively. Any scattering from the embedded active particles is neglected in this analysis. Taking the dosimeter to have dimensions of a 10×10×10 cm^3 cube and considering relevant CCD sizes to be around 3×3 cm^2, the magnification will need to be $m \approx 0.30$. This dosimeter size also yields a maximum depth of $z_{\max} \approx 10$ cm. Further, as the product between solid angle and area is conserved, the f-number of lens 1 will need to be small in order to get the most signal, say about 0.70. Filter, lens transmission, and camera quantum efficiency are set to 0.55, 0.95 and 0.50, respectively. With this,

$$N_{\min} \sim 4 \cdot 10^7 \ \frac{\text{photons}}{\text{Gy} \cdot \text{mm}^3}$$

This number does not only provide information about the required OSL signal from the active material, but also holds information about the concentration of active material needed in the transparent matrix.

Investigation of the OSL signal from commercially available MCP pellets has been undertaken in the group, and these were found to emit $3.3 \cdot 10^8 \ \frac{\text{photons}}{\text{Gy} \cdot \text{mm}^3}$ [9]. With the above result, this implies a minimum concentration of ~10 % MCP powder uniformly dispersed in silicone, for the dosimeter to fulfill the precision criteria. With a concentration this high, refractive index matching of the active particles to the transparent matrix would be essential for scattering to be negligible. This estimated concentration exceeds the result from [9] mainly due to a more conservative estimate of the applied magnification used here. More traditional OSL-active materials, such as Al_2O_3:C, have a higher OSL-efficiency [10], and using these would thus lower the minimally needed concentration of active material.

In summary, this motivates further investigation of the OSL signal from MCP powder embedded in silicone for comparison with the result above and validation of applicability of this material. Such measurements are to be performed in the immediate future.

3. Preliminary 2D readout

As a first step towards a spatially resolved OSL readout, preliminary measurements of commercially available MCP pellets irradiated with 20 Gy have been made using a Falcon Blue 285b-cl EMCCD camera from Raptor Photonics. The pellets were emptied with a laser of wavelength 460 nm transmitted through a circular aperture and imaged onto the pellet with an average intensity of $\sim 750 \frac{\text{mW}}{\text{cm}^2}$. Appropriate filters, resulting in a spectral window between 310 and 400 nm, were used to ensure that only OSL signal was detected. Specifically, three OD4 hard coated filters with cut-off wavelength at 400, 425 and 525 nm from Edmund Optics were used along with two colored glass filters, one from Thorlabs (FGUV5S) and one from Edmund Optics (BG-38). The exposure time of the camera was 2 seconds and the optimal gain setting on the camera using 5-by-5 binning was found to be 2450. The first and last frames from this measurement, corrected for a background contribution, are displayed in Figure 2 in the leftmost and rightmost inset, respectively.

In order to investigate whether the obtained spatially resolved signal had indeed OSL origin, the decay of the spatially resolved signal was extracted by fitting all of the frames with a 2D Gaussian

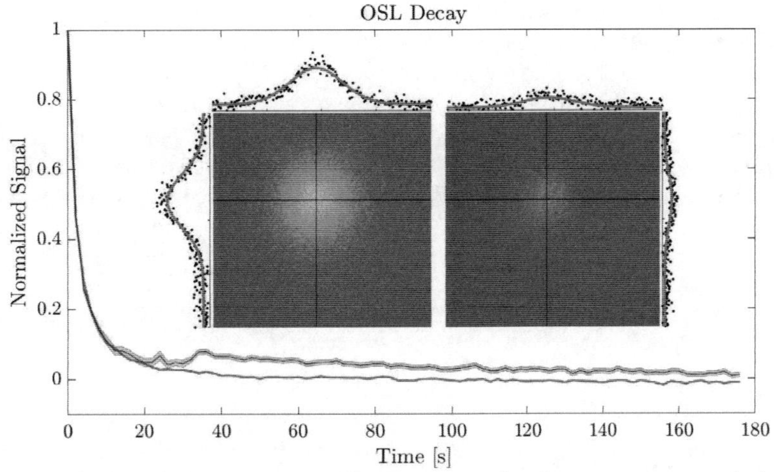

Figure 2. Decay of OSL signal from two pellets measured using a spectrograph (red line) and 2D Gaussian fit amplitudes from 2D measurements (black line) with corresponding fit uncertainties (grey shade). The insets show the background-subtracted 2D-resolved signal from the first (left) and last (right) frame. A slice through the centroid of the 2D Gaussian fit in both directions is displayed for each inset including the fit (red curve). The insets are shown on the same color-scale.

function. Due to a fast decaying signal and a fairly high and fluctuating noise level, a proper Gaussian fit was only possible for the first frame. Amplitudes for the remaining frames were obtained by locking the centroid and width of the Gaussian function to that corresponding to the first frame. Hence, only the amplitude along with the offset and slope of the background plane were fit parameters in all frames except the first. Amplitudes from this analysis can be seen plotted against time in Figure 2.

To compare the obtained decay-signal from the spatially resolved measurement, one of the pellets were investigated in a setup for time-resolved spectral characterization described in [9]. The same laser settings were used to read out this pellet, except with a slightly lower average intensity of $\sim 600 \frac{\mathrm{mW}}{\mathrm{cm}^2}$. This setup allowed for measurement of the pellet's OSL decay by integration over all OSL relevant wavelengths, which is assumed to be representative for all the pellets' decay-behavior when exposed to identical doses. The resulting decay-signal is shown in Figure 2, where the normalized decay-signal from both measurements is seen to be quite similar implying that the spatially resolved data is indeed an OSL signal.

4. Conclusion

We present an estimate of $\sim 4 \cdot 10^7 \frac{\mathrm{photons}}{\mathrm{Gy \cdot mm^3}}$ as the minimum number of photons that an OSL active material embedded in a transparent matrix should emit to be clinically applicable for patient-specific quality assurance. Furthermore, we present 2D spatially resolved measurements from a commercially available MCP pellet. By comparing the decay-pattern of these measurements with that from spectral measurements of the pellets, the signal is attributed to spatially resolved OSL signal. Together, these two results are very promising for the 3D OSL dosimetry concept.

5. Acknowledgements

This research was supported by the Novo Nordisk Foundation.

6. References

[1] Doran SJ 2009 *Appl. Radiat. Isotopes* **67** 393-8
[2] Høye EM, Balling P, Yates E S *et al* 2015 *Phys. Med. Biol* **60** 5557

[3] Høye EM, Skyt P S, Balling P *et al* 2017 *Phys. Med. Bio.* **62** 73-89
[4] D'Errico F, Lazzeri L, Dondi D *et al* 2017 *Radiat. Meas.* **106** 612-7
[5] Taño J, Hayashi S, Hirota S *et al* 2019 *J. Phys. Conf. Ser.* **1305** 012034
[6] Juang T, Adamovics J, Oldham M 2015 *J. Phys. Conf. Ser.* **573** 012039
[7] Sadel M, Høye EM, Skyt PS *et al* 2017 *J. Phys. Conf. Ser.* **847** 12044
[8] Balling P, Høye EM, Muren LP *et al* 2018 *European Patent Application; EP3264137A1*
[9] Nyemann JS, Turtos RM, Julsgaard B *et al* 2020 *Radiat. Meas* **138** 106390
[10] Akselrod MS 2011 *AIP Conf. Proc.* **1345** 274-302

Journal of Physics: Conference Series **2167** (2022) 012034 doi:10.1088/1742-6596/2167/1/012034

Utilizing Pencil Beam Scan Dynamics and a Scintillation Screen to produce 3D Dose Distribution of Proton Beams

Mahbubur Rahman[1], Petr Bruza[1], David J. Gladstone[1,2], Rongxiao Zhang[1, 2], and Brian W. Pogue[1,3]
[1]Thayer School of Engineering, Dartmouth College, Hanover, NH, USA
[2]Dartmouth Hitchcock Medical Center, Lebanon, NH, USA
[3]DoseOptics, LLC, Lebanon, NH, USA

Email: petr.bruza@dartmouth.edu

Abstract. This study describes a method for producing volumetric dose profiles of proton beams from 2D slices of optical scintillation images. The method relies on a high frame rate camera acquisition (100 frames per second), the spot scanning capabilities of current proton pencil beam scanning systems, and a water equivalent scintillation screen. The acquired slices are corrected for optical blurring and ionization quenching and stacked to produce volumetric dose distribution. The volumetric optical dose profile had a pass rate of 98.3% for 2%/2mm local gamma analysis, suggesting the method can accurately measure dose profiles. The method can potentially image all clinical proton beams for pencil beam scanning systems and can extend to imaging patient plans, with further verification.

1. Introduction

In proton therapy, validation of dose distribution is vital to verifying the irradiation system can treat patients and deliver the prescribed plan, particularly due to the significant dose deposition at the Bragg peak. There has been an interest in measuring precise three-dimensional (3D) dose profile of proton beams and scintillation and radioluminescence imaging has shown promising methods of acquiring volumetric dose distributions [1-3]. Optical imaging inherently provides two-dimensional (2D) dose profiles and require reconstruction methods to produce a 3D dose distribution. Some methods include multiple acquisitions of 2D radioluminescence slices of the dose profile [1] or imaging the scintillation projection dose profiles from orthogonal views with multiple cameras [3].

In this study, proton pencil beam scanning (PBS) dynamics produced a 3D dose profile while imaging optical emission from a scintillation screen in a water tank phantom. The imaging method required a single camera acquisition and a single dynamic treatment plan, where the beam scanned across the scintillation screen, for reconstruction of the 3D dose profile. We produced 2D profiles along the crossline of the treatment room from the frames in the acquisition. The optical properties of the scintillation screen and camera detector system (i.e. optical blurring, ionization quenching) were characterized to correct the acquired images. The post-processed dose profile was compared to Monte Carlo (MC) simulated dose profiles to determine the validity of the technique for quality assurance of proton beams.

2. Methods and Materials

2.1. Experimental Set up

We utilized the PBS system (ProBeam, Varian Medical Systems, CA, USA) for imaging dosimetry tests. The phantom, shown in Figure 1a, consisted of an acrylic tank ($15 \times 15 \times 30\ cm^3$) with water and an organic scintillation screen ($27 \times 14.8 \times 0.1\ cm^3$, EJ240G screen, Eljen Technology, TX, USA). The camera imaged the scintillation screen through one face of the tank while the other four faces were covered by black matte plastic. The top face was not included in the phantom to prevent beam broadening or shift in range due to an obstructing slab of material [4]. The scintillation screen was aligned parallel to the imaged face and submerged in water with an edge aligned to the water surface. The phantom was aligned with the centre of the water surface at the isocentre and the scintillation screen along the crossline (XZ) plane. The intensified camera (C-Dose Research, DoseOptics LLC., NH, USA) used a 100mm lens (Canon Inc, Japan) and was positioned 3.25 m away from the isocentre with the optical axis aligned with the isocentre. The camera, capable of high frame rate imaging at 100 frames per second, acquired images of the scintillation in continuous mode with a 10 ms exposure time while a 100 MeV proton beam irradiated the phantom [5]. Cherenkov production in the water from the beam was negligible in comparison to the scintillation intensity from the screen [6,7]. The PBS steering magnet initially positioned the beam along the central axis for imaging of a 2D crossline profile. It then scanned the beam inline (+20 mm to -20 mm, 0.1 mm spot spacing, and 400 MU per spot), while the phantom was kept fixed, for acquisition of optical slices for 3D reconstruction of the dose distribution (described further in Section 2.4). Figure 1b shows example images acquired during acquisition of the scanned beam (and central axis profile).

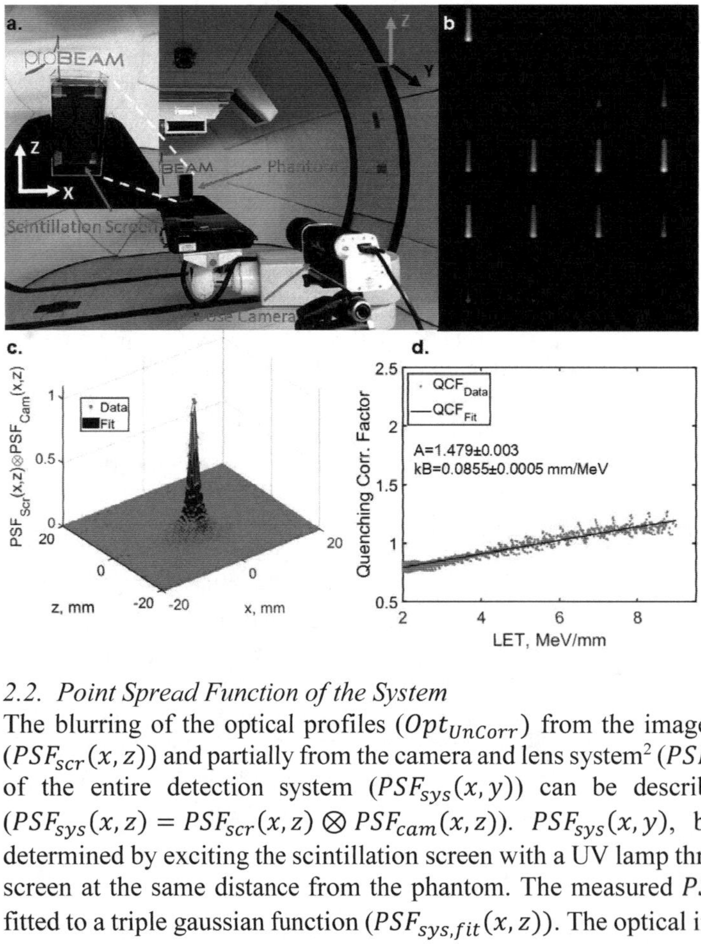

Figure 1. a. Experimental set up for imaging a scintillation screen with the C-Dose camera focused on the scintillation screen along the XZ plane of the isocentre. Camera height and water surface height aligned to height of isocentre. **b.** Montage of example frames acquired while proton beam irradiated at the central axis (top left frame) and scanned axially across the scintillation screen, along the optical axis of the camera. **c.** Point spread function of the camera and scintillation screen detector system **d.** Quenching correction factor data and fit (A, and kB fit values with 95% confidence) from optical images and MC dose

2.2. Point Spread Function of the System

The blurring of the optical profiles (Opt_{UnCorr}) from the images was due to the scintillation screen ($PSF_{scr}(x,z)$) and partially from the camera and lens system[2] ($PSF_{cam}(x,z)$). The point spread function of the entire detection system ($PSF_{sys}(x,y)$) can be described as a convolution of each part ($PSF_{sys}(x,z) = PSF_{scr}(x,z) \otimes PSF_{cam}(x,z)$). $PSF_{sys}(x,y)$, blurring of the entire system, was determined by exciting the scintillation screen with a UV lamp through a small pinhole and imaging the screen at the same distance from the phantom. The measured $PSF_{sys}(x,y)$, shown in Figure 1c, was fitted to a triple gaussian function ($PSF_{sys,fit}(x,z)$). The optical intensity profiles along the central axis and slices from the scan were deconvolved with the $PSF_{sys,fit}(x,z)$ to remove blurring from to the detector system ($Opt_{UnCorr}(x,z) = Opt_{PSFCorr}(x,z) \otimes PSF_{sys,fit}(x,z)$).

2.3. Determine Quenching Parameters

The ionization quenching parameters were determined using the central axis 2D dose distribution profile. The quenching of the scintillations screen response, particularly near the Bragg peak, can be described using Birks' Law [8]:

$$(1) \quad Sc = \frac{A*LET}{1+kB*LET}\phi \propto \frac{(A*D)}{1+kB*LET}$$

Where Sc is the scintillation response, D is dose delivered by radiation source, LET is the linear energy transfer, ϕ is the particle fluence, A is scintillation efficiency, and kB is Birks' or quenching constant. 2D equivalent of the methods described in Wang et al. were implemented to determine the quenching parameters A, and kB [9]. The Monte Carlo (MC) dose (D_{MC}) and LET (LET_{MC}) profile were simulated on the Geant4 based TOPAS MC toolkit, using a validated beam model based on the commissioned beam energies of the Varian ProBeam System [10-13]. The simulations assumed the scintillation screen was water equivalent. The quenching correction factor (QCF) was the ratio of MC dose profile to the measured optical intensity profile. It was defined in a region of interest ($6 \times 9\ mm^2$ ROI) around the Bragg Peak (i.e. region of greatest variation in $LET(x,z)$). The QCF_{data}, shown in Figure 1d, was fitted to

$$(2) \quad QCF_{fit} = \frac{1}{A} + \frac{kB}{A}LET_{MC}$$

The optical image was shifted in the x and z direction to minimize the root-mean-square (rms) error of the fit and the shift with the minimum error provided the quenching parameters ($\delta_z = -1.42\ mm, \delta x = +1.37mm$). The optical profile was corrected for ionization quenching using the fit and the following equation:

$$(3) \quad Opt_{QCorr}(x,z) = Opt_{PSFCorr}(x,z) * QCF_{fit}(x,z)$$

2.4. Reconstruction of 3D Dose Profile

The frames acquired during scanning of the proton beam inline provided a temporal mean intensity profile (from an ROI of $2.5 \times 2.5\ mm^2$ along the central axis at a depth of 20 mm, where LET was approximately constant). The profile was fitted to a Fourier series and the period was determined to be $98.54 \pm .06$ seconds (95% confidence interval). The period and the distance traveled over one scan (40 mm) provided the velocity of the scan, and the number of frames to accumulate for 1mm slices (246 frames) of the dose distribution. Intermittent delay between acquisition of consecutive frames exhibited negligible effect on the measurements [5]. The reconstructed 1mm slices were corrected for optical blurring and ionization quenching, determined in section 2.2 and 2.3, respectively.

3. Results and Discussion

The imaged central axis dose distribution (Figure 2a), after correcting for detector blurring and ionization quenching, matched well with the MC dose distribution with a 2%/2mm local gamma (10% threshold above the maximum dose) pass rate of 99.8%. The central axis depth dose curve, in Figure 2b, agreed relatively well with MC depth dose curve but showed some discrepancy at the entrance of the beam to about 30mm depth (max of 3.2% relative to maximum dose) and fall off at the distal edge of the Bragg peak. The method of scanning the beam across the scintillation screen for acquiring 3D dose distribution was stable because the corresponding frames in future cycles of the scan had a standard deviation of 2.2% (normalized to maximum intensity). The reconstructed 3D dose distribution, shown in Figure 3, agreed well with the MC dose distribution with a 2%/2mm local gamma pass rate of 98.3%.

The imaging method can potentially be improved by considering the optical and physical properties of the phantom. While the density of the scintillation screen is nearly water equivalent (1.023 g/cm^3, 3.33×10^{23} $electrons/cm^3$), the slight discrepancy can lead to a difference in stopping power compared to water and range shift [14]. The shift may be negligible due to the thin screen thickness of

1mm, but a future study will incorporate the properties of the scintillation screen into the MC simulations, to quantify its impact on the proton range. While utilizing image processing methods described in Rahman et al.[5], and correcting with the $PSF_{sys}(x,y)$ addressed optical artefacts from the lens, and camera imaging system, optical distortion from the camera perspective and refraction within the phantom were not considered[2]. At a viewing distance of 3.25m, the 100 MeV beam would result in a shift <0.5mm for the entire profile, but for higher clinically relevant proton energy (e.g. 240 MeV) there can be a shift of ~2 mm potentially requiring a correction for perspective. It is worth noting while this method can resolve volumetric profile for a pencil beam, for complex treatments its potential use would be in providing a 2D dose profile along the central axis.

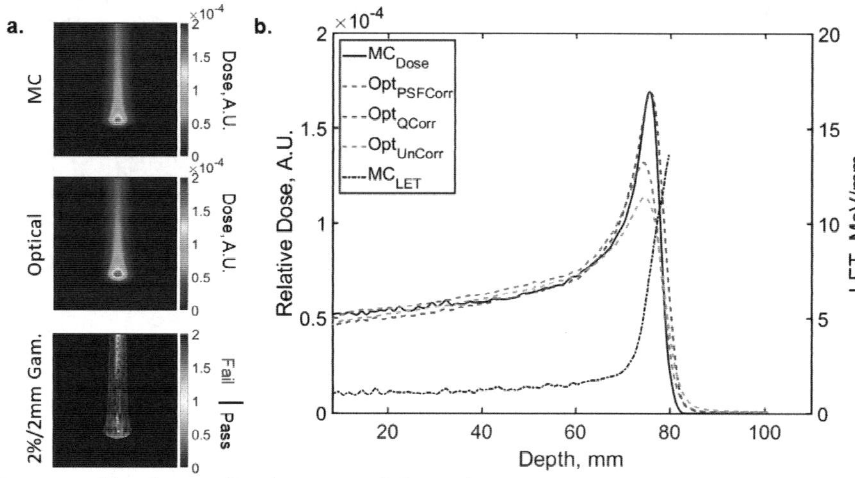

Figure 2. Dose profiles (normalized to sum of dose above 10% threshold of max dose) of the imaged 100 MeV proton beam at the central axis **a.** Crossline quenching corrected optically measured dose distribution and 2%2mm local gamma analysis (10% threshold above max dose) in comparison to the MC dose. **b.** Central Axis depth dose curves (MC simulated dose, and optical with no corrections, PSF correction, and quenching correction) and central axis MC LET.

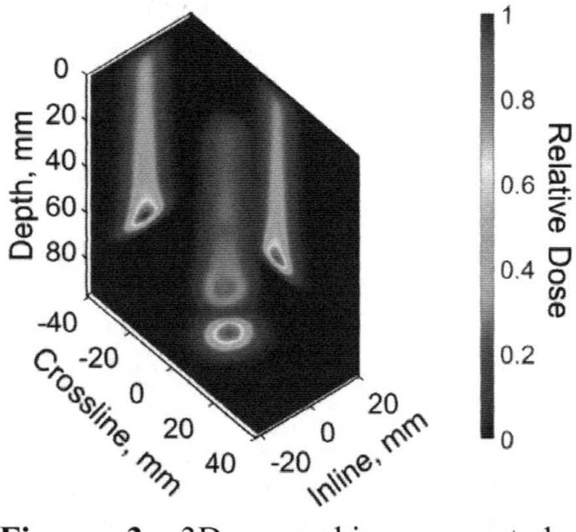

Figure 3. 3D quenching corrected optical dose distribution of 100 MeV beam using the scanning PBS. Dose distribution along the depth plane at the Bragg Peak, inline plane through the isocentre, and crossline plane through the isocentre are included.

4. Conclusion

A novel method was described here that utilized the PBS dynamics to reconstruct an accurate 3D dose distribution of clinical proton beams. A resolution of 1 mm^3 was achieved and the dose distributions (both central axis and 3D) agreed with the MC dose distribution with a 2%/2mm local gamma pass rate of above 98% The central axis depth dose curve's discrepancy at the entrance of the beam and at the distal edge of the Bragg peak need to be explored. The method assumed that the steering magnets scanning the proton beam inline over 40 mm do not significantly perturb the beam shape, which needs to be verified. The imaging technique needs to be verified for accuracy across the clinical proton beam energy range. This method provides fast acquisition of the volumetric dose profile and is dictated by the time of a single scan across the scintillation screen. The scintillation screen can potentially image central axis profiles of dynamic treatment plans and will require further validation (i.e. spread out Bragg peak).

5. Acknowledgments
This project has been sponsored by National Institutes of Health research grants R01 EB023909. The authors acknowledge the Irradiation Shared Resource at the Norris Cotton Cancer Center at Dartmouth with NCI Cancer Center Support Grant 5P30 CA023108-41.

6. Disclosures
Brian W Pogue reports financial interest in DoseOptics LLC, a company developing cameras and software for the use of Cherenkov imaging in radiotherapy dosimetry.

7. References
[1] Horita R, Yamamoto S, Yogo K *et al* 2019 *Biomed. Phys. Eng. Express* **5** 045033
[2] Robertson D, Hui C, Archambault L *et al* 2014 *Phys. Med. Biol.* 59 23-42
[3] Darne CD, Alsanea F, Robertson DG *et al* 2017 *Phys. Med. Biol.* **62** 5652-67
[4] Schaffner B 2008 *Phys. Med. Biol.* **53** 1545-62
[5] Rahman M, Brůža P, Langen KM *et al* 2020 *Phys. Med. Biol.* **65** 165014
[6] Glaser AK, Zhang R, Gladstone DJ *et al* 2014 *Phys. Med. Biol.* **59** 3789-811
[7] Robertson D, Mirkovic D, Sahoo N *et al* 2013 *Phys. Med. Biol.* **58** 261-73
[8] Birks JB 1964 *The Theory and Practice of Scintillation Counting*. 1st ed. Pergamon
[9] Wang LLW, Perles LA, Archambault L *et al* 2012 *Phys. Med. Biol.* **57** 7767-81
[10] Langner UW, Eley JG, Dong L *et al* 2017 *J. Appl. Clin. Med. Phys.* **18** 96-107
[11] Chang C, Huang S, Harms J *et al* 2020 *Med. Phys.* Published online January 2020
[12] Huang S, Kang M, Souris K *et al* 2018 *J. Appl. Clin. Med. Phys.* **19** 558-72
[13] Rahman M, Bruza P, Lin Y 2020 *Med. Phys.* Published online October 2020
[14] Scott Ingram W, Robertson D, Beddar S 2015 *Nucl. Instrum. Methods Phys. Res. Sect. Accel. Spectrometers Detect. Assoc. Equip.* **776** 15-20

AUTHOR INDEX

Institute of Physics
Dirac House, Temple Back
Bristol BS1 6BE UK

ISSN: 1742-6588
ISBN 978-1-7138-5138-7

CURRAN ASSOCIATES INC.
proceedings.com